SpringerBriefs in Environmental Science

For further volumes:
http://www.springer.com/series/8868

Mikhail Butusov • Arne Jernelöv

Phosphorus

An Element that could have
been called Lucifer

 Springer

Mikhail Butusov
The International Center for Advanced
and Comparative EU-Russia Research
Vienna, Austria

Arne Jernelöv
Institute for Futures Studies
Stockholm, Sweden

ISSN 2191-5547   ISSN 2191-5555 (electronic)
ISBN 978-1-4614-6802-8   ISBN 978-1-4614-6803-5 (eBook)
DOI 10.1007/978-1-4614-6803-5
Springer New York Heidelberg Dordrecht London

Library of Congress Control Number: 2013934218

Printed on acid-free paper

Springer is part of Springer Science+Business Media (www.springer.com)

Phosphorus: An Element That Could Have Been Called Lucifer

Mikhail Butusov and Arne Jernelöv

Introduction

Is Phosphorus a Sustainability Bottleneck?

In 1669 the German alchemist Henning Brand was the first to produce and identify phosphorus as an element. In his search for gold he heated a large volume of human urine to evaporate the water. In the very end he overheated the condensate somewhat. When he carefully checked the remains for traces of gold, he noticed a faint glow. It came from an element he did not know. He had produced white phosphorus, which emitted some light when it was oxidized in air. He initially called it the bearer of light. According to the tradition at the time, he could have used Greek or Latin as the basis for the scientific name. He opted for Greek, and light bearer became Phosphorus. Had he opted for Latin, it would have been Lucifer. Given what might await mankind, if we run out of the element as fertilizer in agriculture, Lucifer might have been a more appropriate choice.

Our book starts with depiction of the role of phosphorus in the creation of life and its evolution. Then it outlines in which vital biological processes different phosphates participate to sustain the life of all flora and fauna, from DNA molecules through building up body tissues. The most crucial function of phosphates is in agriculture. This role was noticed long ago, but only in the nineteenth century did the discovery of mineral fertilizers make it possible to sustain the needs of the growing global population, thus initiating a "green revolution." For many decades, the complexity of the interactions "fertilizer–soil–plant roots" was underestimated, thus causing massive environmental and economic damage, such as intensive mining of the depleted natural phosphate reserves that continues worldwide, decay of the natural soil structure, land desertification, and eutrophication of waters. The lessons of what happened, on a much lesser scale, before the nineteenth century because of the scarcity of phosphates, have been forgotten during recent decades. In the meantime, the production of natural phosphates reached its peak a few years ago. What are the potential consequences? In our opinion, to help avoid a forthcoming global disaster, immediate implementation of phosphate recycling technologies, for example, from municipal wastes, should be fostered.

Contents

Chapter 1
The Role of Phosphorus in the Origin of Life and in Evolution

Phosphorus, in the form of phosphate, has played an important role in the origin and evolution of life on several different levels. It was, most likely, a key component in the early precursors of RNA that existed before real life, where it both stored information and acted as a catalyst. It plays an essential role in both the genetics and the energy systems of all living cells as well as in the cell membrane of all modern cells. Phosphorus has also had a decisive role in forming the climatic and atmospheric conditions that set the boundary conditions for evolution and led to us humans and the world we know now.

1.1 How Many Origins of Life?

A very central question with regard to life on Earth is whether all life forms have one common origin or whether life originated independently several times and possibly through more than one route. Most scientists in the field today share the view that all known life forms have one common ancestor: that is, even if life evolved more than once, only one of those occasions led to all the living organisms present on Earth today.

One of the most central criteria used to define life is the ability to metabolize, to convert molecules absorbed from the surroundings to building blocks of its own structures. Another is the capacity to self-reproduce, for an organism to build copies of itself. Both these functions, and many others, require an energy system in which energy can be stored, transported, and released in a controlled and timely fashion.

The compelling argument for the one origin of life theory is the uniformity of the genetic system based on the nucleic acids DNA and RNA and the energy system based on ATP known among all existing organisms. The likelihood that such complicated systems would have evolved twice and in parallel seems very slim, thus suggesting one origin of all life forms.

M. Butusov and A. Jernelöv, *Phosphorus: An Element that could have been called Lucifer*, SpringerBriefs in Environmental Science 9, DOI 10.1007/978-1-4614-6803-5_1, © Mikhail Butusov and Arne Jernelöv, 2013

1.2 Nucleic Acids, Phosphorus, and the Road to Life

DNA and RNA (deoxyribonucleic and ribonucleic acid) are made up of a long chain of nucleotides, each consisting of a nitrogenous base, a ribose sugar, and a phosphate group (Fig. 1.1.)

The crucial energy molecule ATP (adenosine triphosphate) consists of an adenine base attached to the sugar ribose and three phosphate groups (Fig. 1.2).

The thing to note here, before getting bewildered by the biochemistry, is the presence of the phosphate groups in both systems. Thus, phosphorus as phosphate is part of the most basic molecules in all life and is, in the most literal meaning, an essential element.

Francis Crick, the man who in 1962 shared the Nobel Prize with James Watson for the discovery of the structure of DNA in 1953 (Watson and Crick 1953), formulated what became known as the central dogma of molecular biology and described the roles of DNA, RNA, and proteins in the living cell. Basically, the theory states that DNA stores the genetic information, and RNA transports it to the centers of

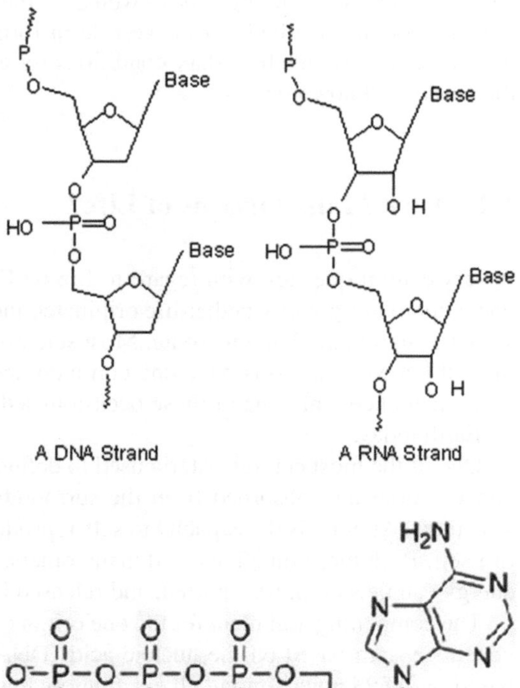

Fig. 1.1 Structures of molecular strands of DNA and RNA

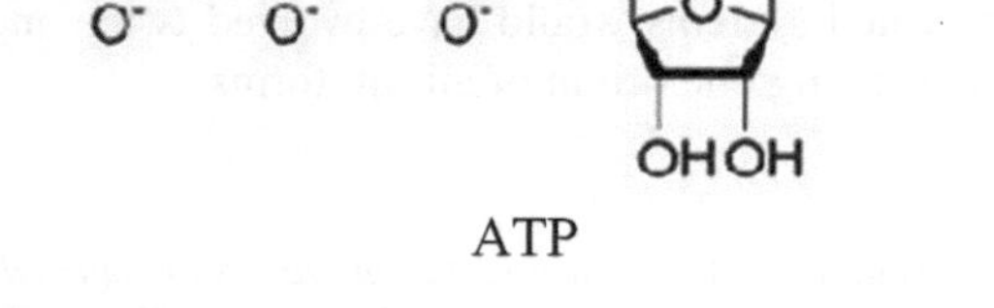

A DNA Strand   A RNA Strand

Fig. 1.2 Structure of the adenosine triphosphate (ATP) molecule

ATP

protein synthesis, where it directs the formation of proteins which then "do the job" (Crick 1970). In certain circumstances information can pass back from RNA to DNA, but never from the proteins back to the nucleic acids.

With this dogma as the starting point, to figure out the organic chemical steps that led to life and biochemistry becomes a 'chicken-and-egg exercise'.[1] Which came first: the proteins that build the cell structures and catalyze the chemical reaction in the cell, the DNA which contains the information about how the proteins should be built, or the RNA that transports the information from the place of storage to the place of use?

Another Nobel laureate, Jack Szostac, who was awarded his 1991 prize for discoveries relating to the structure and function of telomeres and the ends of chromosomes, resolved this apparent dilemma by emphasizing the likely simplicity of early life and focusing his current research on the origin of life on fatty acids and nucleic acids of the RNA type (Szostac 2012).

It was known, long since, that RNA, although chemically much simpler than DNA and much less effective, also can store information. Furthermore, Tom Cech (1990) and Sidney Altman et al. (1989) independently demonstrated that RNA has some catalytic properties and can perform some functions previously thought to be the exclusive domain of proteins, which gave them the Nobel Prize. Since then RNA has assumed the role as the likely predecessor of both DNA and proteins in cellular functions. Jack Szostac and his coworkers have used in vitro selection to isolate and characterize a large number of simple nucleic acids and sequences with specific properties regarding catalysis and ligand binding. They have also used it as a tool to isolate functional nucleic acid and protein molecules from a large pool of random sequences.

As a result, in a pre-biotic world with simple organic acids and nucleotides present, the following first physicochemical steps toward a living cell have been postulated.

1. Fatty acids with a hydrophobic and a hydrophilic end lump together and form vesicles. When they encounter free fatty acids in solution, they incorporate them, and when two of such structures collide, they merge. Thus, the vesicles grow. These processes are driven by physics (thermodynamics).
2. At a certain size, dependent on, for example, type of fatty acid, pH, and temperature, the vesicle adopts a tubular branched form. Parts of it are easily broken off by such forces as water turbulence.
   Through these processes of "eating," growing, and multiplying, units of fatty acids are formed.
3. Among the several hundred types of nucleotides available and, at one time or another, incorporated in fatty acid vesicles, some can polymerize under prevalent conditions. Phosphoramidites, a group frequently used today in the synthesis of RNA and DNA, belong to them.

---

[1] In the context of evolution, the 'chicken and egg' metaphor is actually a dumb one as the question which came first has an obvious answer. Wherever we draw the evolutionary line for when the chicken as a species first appeared, the egg was there long before as more primitive birds, dinosaurs, lizards, and fishes all have eggs.

These polymers have "walls" that are easily permeable for small organic molecules, and as such will be incorporated and maintained because of higher solubility in fats than in water and the arrangement of the hydrophobic and hydrophilic ends of the fatty acids.

4. Monomers will base pair with single-stranded templates and self-ligate.
5. Nucleic acids with catalytic properties such as modern RNA can convert sib-molecules to their own type and thereby accelerate the process.
6. By multiplying and polymerizing, nucleotides increase the osmotic pressure inside the fatty acid vesicle and cause it to expand, thereby increasing the speed with which it incorporates more fatty acids, grows, and divide.

Thus, there is now a system with "eating," growing, multiplying vesicles that contain polymerizing and self-reproducing nucleic acids, which could contain information, and that can catalyze some chemical processes. That system may be all it takes for natural selection and a Darwinian evolution based on competition between vesicles or cells to begin.

For the purposes of this book, it is noteworthy that phosphorus according to this model plays an important role from the very onset of life—or even before. The role will become even greater later, when the fatty acids that in the model form the "walls" of the pre-cells are replaced by phospholipids and ATP acquires its crucial role in the energy system of the cells.

1.3 Where Did It All Happen?

In the foregoing brief description of the evolutionary model, one important element is missing: geography. Where would these first steps toward life have taken place? Early researchers thought of this process happening in shallow waters at the Earth's surface in an atmosphere still devoid of oxygen and with lightning as the energy source for production of the first organic molecules (Horgan 1991).

Later, it was discovered that there is an abundance of organic molecules in space, which gave support to an old hypothesis of the 1903 Nobel laureate Svante Arrhenius stating that life came to Earth as spores from space (Arrhenius 1908).

Later still, deep-sea vents, where hot mineral-rich water from inside the Earth's crusts gushes into the cold bottom waters of the oceans, have appeared as a front-runner candidate of possible locations for the origin of life on Earth. Here the originator was not a Nobel laureate (although there are those who thinks he deserves a Nobel Prize), but a much younger and less-established scientist, Jack Corliss (Corliss et al. 1981). An important feature here are convection currents, which move water with dissolved components and suspended particles between hot zones adjacent to the vent and the colder zones away from the vents. This transport allows for a process called thermo-synthesis, which in turn can yield the self-organization and dissipative structures that are required in any model of the chemical origin of life. The thermal cycling in germination and cell division can be seen as a relict of primordial thermo-synthesis.

In addition to a steep temperature gradient, the vents also offer an environment with high concentrations of dissolved and suspended material, where inorganic compounds of such elements as iron, sulfur, nitrogen, and hydrogen, which in different oxidation stages can act as electron donors and acceptors, result in energy release and the possibility for chemosynthesis.

1.4   First Life and a Second Step

The first living cells—a stage that was reached some 4 billion years ago—must have had similarities with today's prokaryotes, primitive life forms of which two large groups still exist, Bacteria and Archaea. They lack a cell nucleus or any membrane-encased organelles, and their genetic material is organized in a single loop. In the more advanced eukaryotes the genes are arranged in chromosomes, which are enclosed in a membrane-bound cell nucleus, and they possess membrane-bound organelles.

The step from prokaryotes to eukaryotes is one of the evolutionary leaps. It is supposed to have happened some 2 billion years ago, at approximately the same time as cell respiration and an organized energy storage and distribution system within the cell developed.

This stage was predated, however, by another really fundamental change in the living conditions for practically all types of life that started without much ado perhaps 3.5 billion years ago. Some organisms, with similarities to today's cyanobacteria, developed photosynthetic abilities and learned to make organic compounds from inorganic compounds with light as the energy source. As a by-product, they produced oxygen. Initially, the produced oxygen quickly reacted with reduced forms of iron, sulfur, and other elements, but as oxygen production continued those electron donors became more sparse and free oxygen became present in the atmosphere and dissolved in the ocean and other waters. This change happened some 2.5 billion years ago.

1.5   The Worst Environmental Catastrophe Ever

That fundamental change in the conditions for life on Earth triggered the largest ecological catastrophe in our planet's history, wiping out practically all existing life forms. It has been estimated that more than 99 % of the organisms existing then died out. It could be that this event exterminated life of other origins than the one and only that we find today, but this is impossible to know. Toxic as it was to living cells, oxygen permitted the development of higher organized life because, through oxidation of organic matter, it gave the survivors access to more cell energy than was available to previous life forms.

A relevant question in this context is why oxygen was that toxic to the early life forms when their primitive relatives today tolerate it. One answer is in the type of fatty acids that form the cell membrane. Pre-oxygen organisms had forms that turned rancid in contact with oxygen, whereas the phospholipids of later forms were far less vulnerable to this form of decay.

Another big step, partly integrated with the new arrangements of the cells, is the step from single-celled to multi-celled organisms. Here it is believed that early eukaryotic cells came to integrate with primitive bacteria, which with time became the mitochondria. Later, some organisms integrated cyanobacteria that became chloroplasts. The latter happened at multiple occasions at different times: the first appears to have been a single event. On a timeline, multicellular algae appeared 1.3 billion years ago, whereas their animal counterparts came much later, a mere 600 million years ago. Their closest living relatives are the sponges. Another early group was the Eandromeda, with an as yet unclear phylogenetic position.

1.6   The Role of Phosphorus

So, where in this development from first life to multicellular organisms does phosphorus come in? In biochemistry and physiology, practically everywhere. As already said it is an essential part of the first RNA-like nuclides, it is a central part of the phospholipids that quickly replaced the fatty acids from pre-life as cell membranes, and as a key component in ATP it participates in cell respiration as well as in chemosynthesis and photosynthesis.

Phosphorus, however, is also a key element for the evolution of life in that it played an important role for the processes resulting in an atmosphere with oxygen as one of the major components, the drastic consequences of which were already touched upon (this can be seen as the first manifestation of the luciferous character of phosphorus). The global drama associated with and giving rise to this change may be worth a short tale of its own.

1.7   Changing Atmosphere and Changing Temperatures

The early atmosphere of planet Earth had no free oxygen and thus no ozone layer that absorbed radiation. Methane and carbon dioxide were present in high concentrations and, as they are potent greenhouse gases, the planet was warm.

In the oceans, the concentrations of many ions including phosphate and reduced forms of iron and manganese and sulfur were much higher than they are today, and the water was of course anoxic.

In this environment, the relatives of present-day cyanobacteria acquired a type of chlorophyll and embarked on a phototrophic life path. Initially, photosynthetic bacteria probably did not produce molecular oxygen but used reduced sulfur as the electron donor. Later, early cyanobacteria "learned" to oxidize water, thereby producing the highly toxic free oxygen as a by-product to organic matter. For almost a billion years this led to little more than oxidation of the aforementioned reduced forms of iron, manganese, and sulfur, for example. However, as iron and manganese are reduced, their solubility decreases, and instead of being dissolved in the water mass, they precipitated as deposits on the sea floor. In the precipitation process, phosphates were also included. Both iron and phosphate were and are essential elements for the cyanobacteria themselves, in the cytochromes, for example, and in iron–sulfur proteins and phosphorus as indicated earlier.

With lower concentrations in the water mass, phosphates became limiting factors for the growth and photosynthetic activities of the cyanobacteria themselves.[2] The question of nitrogen is different. Modern cyanobacteria have the capacity to fixate atmospheric nitrogen and turn it into usable nitrate or ammonia, whereby they produce their own nitrogen fertilizer and are not dependent on it being present in ionic form in the water. It is not well known when they developed this ability, but most scholars hold it that it was early on.

Before the onset of life, carbon dioxide levels in the atmosphere had been high, but as the cyanobacteria continued to convert it to organic matter, atmospheric levels slowly decreased over the millennia. Lower concentrations of carbon dioxide in the atmosphere and in the water contributed to reduced production of organic matter and oxygen, but compared to the change in availability of iron and phosphorus, this was a very gradual process. By the time when carbon fixation and oxygen production by cyanobacteria took a real dive, there was already a fair amount of free oxygen in the atmosphere, and much of the methane gas that had been there earlier had been oxidized. Methane gas is a very powerful greenhouse gas, even much more so than carbon dioxide, and less methane—and less carbon dioxide—in the atmosphere meant that the Earth started cooling.

1.8   The Huronian Ice Age

This is a likely factor behind the first known Ice Age, which is called the Huronian as the first evidence of it was found around Lake Huron in present-day North America. It occurred from 2.4 to 2.1 billion years ago, and its traces have been found not only in North America, Europe, and Asia but also in Africa.[3]

---

[2] It has been argued that the co-precipitation of oxidized iron and phosphates with which we are familiar today would not have occurred the same way in the ancient oceans because the concentration of silica was much higher then and that silica would have outcompeted phosphorus on the iron-binding sites, and the phosphate therefore would not have become a limiting factor for photosynthesis at the time. However, a lower rate of carbon burial strongly points to a lower carbon fixation and the iron–phosphate precipitation mechanism provides a reasonable explanation for it.

[3] It should be noted that more than 2 billion years ago the continents were not in the same positions as they are today. Over the history of the Earth there have been at least four occasions when all the landmass formed one supercontinent surrounded by one ocean. The land then broke up into separate continents and reformed again. The latest supercontinent, called Pangaea, was formed 300 million years ago and broke up to form the globe with which we are familiar some 200 million years ago. The oldest known supercontinent, Columbia or Nuna, was probably assembled some 2 billion years ago, that is, at the end of or just after the Huronian glaciation. It lasted for about 500 million years, but the exact configuration and geodynamic history are not well understood. It was followed by Rodinia (1,100–750 million years ago), the short-lived Pannotia (600–540 million years ago) with large amounts of land near the poles and a small equatorial strip connecting the polar masses, and Pangaea.

The Huronian Ice Age was a long one, apparently one with some interglacial periods, and it lasted in total for some 400 million years. What brought it to an end is not very clear. Scientists have suggested such causes as a change in Earth's tilt; increased volcanic activity that spewed carbon dioxide into the atmosphere; increased solar activity with more radiation reaching Earth; continental shifts that brought more of the landmass to the tropics; and many other processes which, isolated or in tandem, could start heating the planet after the long cold spell. Whatever was the trigger, once the process began, the photosynthetic activities of the cyanobacteria accelerated as the melting ice scraped the land and brought great amounts of minerals and salts into the ocean, including limiting nutrients and trace elements such as iron and phosphorus.

The massive ice sheets that fell into the oceans and formed gigantic icebergs cooled the surface waters even as the planet was warming and changed the oceanic current systems, leading to new large upwelling zones, where nutrient-rich bottom water came up into the light at the surface.

Bioproduction jumped, and oxygen became a permanent component in air at a concentration of approximately 1 %.

1.9   And On It Goes

The tectonic plates, and the continents on top of some of them, continued their movements, the supercontinents were formed and broken up, and the biological evolution went on to create the eukaryotic cell. Some of these cells incorporated cyanobacteria, which evolved to become chloroplasts within the host cells. The first multicellular algae appeared, but the overall rate of photosynthesis does not seem to have changed drastically. Perhaps the gradual winding down of the fertilizing effect of the melting Huronian ice was roughly compensated by the growing role, as primary producers, of the more efficient algae as compared to cyanobacteria.

1.10   The Cryogenian Ice Age

Around 800 million years ago it was time for another major ice age, the Cryogenian one sometimes also called the Sturtian-Varangian Ice Age. In all likelihood this was the coldest period in Earth history, and in popular literature it is often referred to as 'snowball Earth.' There is some controversy as to the full extent of the ice cover. It is clear that significant spurs of it can be seen on all continents, and most scientists in the field consider it likely that practically all land was ice covered, with the exception of the highest mountains and some areas with extremely low precipitation, much like Antarctica today. Some even think that the oceans were ice covered, which would have meant that Earth from space at the time would have looked like a gigantic snowball.

What triggered the onset of the Cryogenian Ice Age is much debated, although some modeling work highlighting the position of the then supercontinent Rodinia, with the landmass close to the South Pole, seems to offer a plausible explanation.

The position of the continent prevented the normal heat exchange between equatorial and polar regions of the globe, resulting initially in larger temperature differences than normally occur between them. This disparity triggered the icing of the polar regions, including much of the landmass of Rodinia. With snow and ice there, the albedo, the light reflectance, increased, and a significant part of the sunlight that reached Earth was reflected back into space instead of heating the planet. As more and more of the ocean surface became ice covered, this process of albedo increase became more and more important.

The ice cover of the oceans also had another important effect that contributed to the cooling: water evaporation decreased, leading to a sharp reduction in humidity and precipitation. Water vapor is in itself an important greenhouse gas, although we seldom mention it in the current debates on global warming, so drier air means a colder atmosphere. The sharp reduction in precipitation is one of the reasons why some scientists doubt that the Rodinia supercontinent was fully ice covered. There would not be enough snowfall to replace the ice that unavoidably glides toward the sea, they argue. Large areas would have to be bare, cold rocks.

Lack of precipitation would also affect the global carbon cycle as there would be little rain to wash out carbon dioxide from the atmosphere and, with much less light reaching the ocean waters under the ice, photosynthesis of algae is substantially reduced, as in the present-day Arctic Ocean. Similarly, the lack of rainfall meant that hardly any runoff water reached the ocean and thereby the flow of eroded sediments largely stopped (Fig. 1.3).

One would expect that all this should have had a major effect on life forms and evolution, but sedimentary records do not consistently show this. There are reports about dips in both abundance and biodiversity during the Cryogenian Ice Age, but then there are others that show no significant change. Several groups appear for the first time in the fossils from the period, including red algae, dinoflagellates, ciliates, and testate amoebae. The latter are especially significant from an evolutionary perspective as they are the first fossil evidence of heterotrophic eukaryotes.

When it comes to the reason for the end of the Cryogenian era, the experts are in good agreement. As a consequence of volcanic activity, the concentrations of carbon dioxide in the atmosphere became very high during the 200 million years the Ice Age lasted, helped no doubt also by the low washout by precipitation and the reduced photosynthetic activity by the light-starved algae and cyanobacteria. The carbon dioxide caused the atmosphere to heat up, the ice started to melt, the albedo decreased, water evaporation from ice-free sea areas increased, and the whole set of processes that triggered the onset of the ice age went into reverse and caused its end. In addition, the Rodinia supercontinent started to break up, and heat transfer from the tropics to the poles normalized.

Similar to the situation after the Huronian Ice Age, melting ice carried vast amounts of minerals and nutrients, including phosphorus, to the oceans and photosynthesis soured. The oxygen level in the atmosphere jumped to some 10 %, an

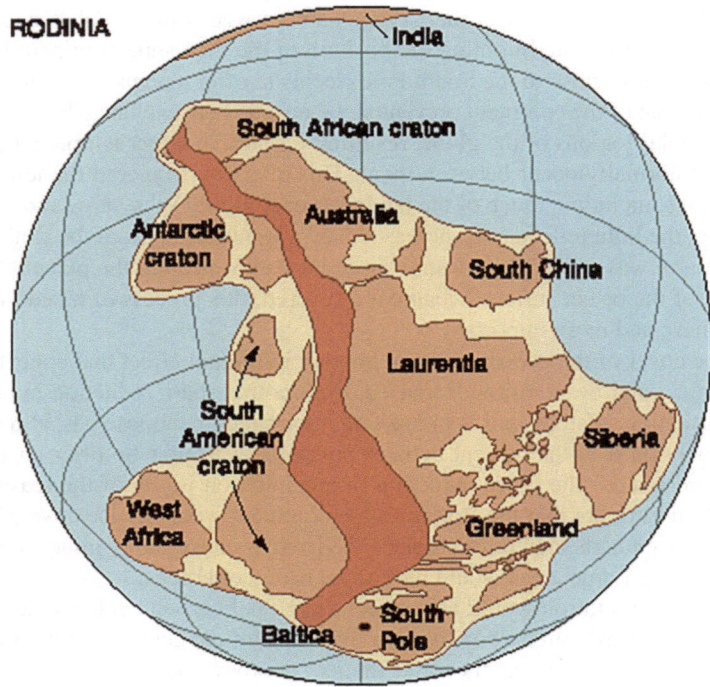

Fig. 1.3 Rodinia: the supercontinent that existed before Pangaea (NASA database)

ozone layer formed, and Earth's surface was protected from much of the sun's high-energy radiation.

The higher oxygen levels also in seawater allowed the development of multicellular animals and paved the way for the so-called Cambrian explosion, a period of very fast evolution and diversification of plant and animal life forms that started 540 million years ago. The radiation shield of the ozone layer allowed plants and animals to colonize the land, and the globe began to look like the one with which we are familiar.

1.11   Was This Really the Chain of Events?

There are, however, a few twists in all of this that may be worth some further discussion.

First, there is this time overlap between the ice ages and the rapid increases of the oxygen content of Earth's atmosphere, and the mechanisms as described with phosphate in a central role are highly plausible. The twist to this cause of events was that the rise in oxygen levels, on closer scrutiny of the available records, seemed to have started *during* the ice ages and not merely at their very end or after them.

The explanation could well be that most ice ages, and certainly the two described here, were not just 100 millions of cold years with ever-growing inland and ocean ice sheets. Instead, there were a number of interglacial periods, when ice melted away and minerals and nutrients were flushed out to sea, before the freezing resumed again. Repeated such episodes during the ice age could account for the rise in atmospheric oxygen *during* the period. So far, however, the time resolution of our observations available for those past periods does not allow either verification or falsification of this explanation.

Second, looking at polar regions today, there is an apparent explosion of plankton algae following an ice melt. A large part of these start their spring growth attached to the ice either on the surface of inland ice or on the underside of flowing ice: in the latter case, much like algae on the glass walls of an aquarium. When the ice melts, the algae turn from a surface-attached life to an existence as free-floating plankton. So far this explanation for the photosynthesis explosion associated with the early ice ages has been rejected on the grounds that without oxygen in the atmosphere and a stratospheric ozone layer, radiation at the Earth's surface would be too strong for cyanobacteria and algae, for example, to survive on the ice sheets. However, it has also been suggested that cyanobacteria in the then chemical environment would naturally precipitate iron–silica minerals that would coat them and give them an effective shield against UV radiation.

1.12   A Very Short Summary

The content of this chapter could be summarized in saying that phosphorus is an essential element both for the origin of life and for its further evolution by being part of fundamental molecules that regulate such functions as genetic mechanisms, cell membranes, and the cell energy systems of all known living organisms. Phosphorus also played a major ecological role in bringing about the Earth as we know it and the life we have on the planet by fertilizing the photosynthetic bacteria and algae that gave us oxygen.

1.13   May Be Not Totally Essential After All

Phosphorus shares this status, of being part of all known life, with five other elements: carbon, oxygen, hydrogen, nitrogen, and sulfur. In discussing the possibilities for life on other planets, scientists have speculated about the possibilities of life based on some other elements. Silica has been suggested as an alternative to carbon, sulfur as a replacement for oxygen, and arsenic for phosphorus. F. Wolfe-Simon, a geo-microbiologist at the NASA Astrobiology Institute, decided to see if any indications of arsenic replacing phosphorus were to be found among bacteria living in high-arsenic environments on Earth. The results of her study were presented in *Science* online in December 2010 and caused some stir among fellow biologists

(Wolfe-Simon et al. 2011). She and her colleagues sampled bacteria from the sediments of a California lake with extremely high levels of arsenic and cultivated them in media with ever-lower levels of phosphate and higher and higher concentrations of arsenic. One type of bacteria, a member of the *Halomonadoceae* family of Proteobacteria, survived well and kept reproducing also when the medium was virtually phosphorus free, indicating that it could indeed use arsenic instead of phosphorus for most if not all essential functions. Chemical analysis also found arsenic deeply embedded in the bacterial DNA, and radioactive arsenic, when supplied, turned up in cell membranes, proteins, lipids, ATP, and glucose in concentrations similar to those of phosphate in normal cells.

References

Altman S, Baer MF, Bartkiewicz M, Gold H, Guerrier-Takada C, Kirsebom LA, Lumelsky N, Peck K (1989) Catalysis by the RNA subunit of RNase P: a minireview. Gene 82(1):63–64

Arrhenius S (1908) Worlds in the making: the evolution of the universe. Harper and Row, New York

Cech TR (1990) Self-splicing of group I introns. Annu Rev Biochem 59:543–568

Corliss JB, Baross JA, Hoffman SE (1981) An hypothesis concerning the relationship between submarine hot springs and the origin of life on Earth. Oceanol Acta, p 59. Proceedings of the 26th International Geological Congress

Crick F (1970) Central dogma of molecular biology. Nature 227(5258):561–563

Horgan J (1991) "In the beginning…". Scientific American. February :100–109. http://www.scienceclarified.com/dispute/Vol-1/Did-life-on-Earth-begin-in-the-little-warm-pond.html#ixzz2CHsxgRdW. Accessed 15 Nov 2012

Szostac J (2012) http://exploringorigins.org/. Accessed 15 Nov 2012

Watson JD, Crick FHC (1953) A structure for deoxyribose nucleic acid. Nature 171:737–738

Wolfe-Simon F, Blum JS, Kulp TR, Gordon GW, Hoeft SE, Pett-Ridge J, Stolz JF, Webb SM, Weber PK, Davies PCW, Anbar AD, Oremland RS (2011) A bacterium that can grow by using arsenic instead of phosphorus. Science 332(6034):1163–1166 (published online 2 Dec 2010)

NASA database http://rst.gsfc.nasa.gov/Sect19/Sect19_2a.html. Accessed 15 Oct 2012

Chapter 2
Phosphorus in the Organic Life: Cells, Tissues, Organisms

As already mentioned (see Chap. 1), in the living cell phosphorus plays a decisive role in three different essential structures:

- In the cell membrane
- In the storage and retrieval system for genetic information, DNA and RNA
- In the energy system, ATP

Furthermore, in vertebrate animals it is an important component in sinew, cartilage, bone, and enamel.

2.1 The Cell Membrane

The fundamental building block of all living organisms is the cell. Some organisms consist of just one cell whereas others are built up by hundreds of millions of cells. All cells are separated from the surroundings by an envelope—the cell membrane. This membrane, a totally necessary structure, allows a much higher concentration of water-soluble compounds and water-suspended particles inside than outside the cell. The cell membrane consists of chains of fatty acids, the molecules of which contain 16 to 20 carbon atoms and a phosphate group at the end: the so-called phospholipids. In eukaryotic cells, there are also sugar molecules and membrane proteins attached to the phospholipids: these are arranged in double layers with the water-soluble phosphates pointing both inward and outward, with the tails of the fatty acids making up the middle (Singer and Nicolson 1972). The direct function of the phosphate group is to provide the essential orientation of the phospholipids, which in turn gives the cell membrane its fundamental characteristics.

M. Butusov and A. Jernelöv, *Phosphorus: An Element that could have been called Lucifer*, 13
SpringerBriefs in Environmental Science 9, DOI 10.1007/978-1-4614-6803-5_2,
© Mikhail Butusov and Arne Jernelöv, 2013

2.2   DNA and RNA

Watson and Crick (see the reference list of Chap. 1) in 1953 presented the structure of DNA as a double helix (Fig. 2.1), and were awarded the Nobel Prize in medicine for this discovery 9 years later. However, DNA had been found much earlier, in 1868, when a Swiss medical doctor and biologist, Friedrich Miescher, found a phosphorus-containing compound in the nucleus of cells (Dahm 2005). The compound, he noted, was of a completely new type and very different from proteins, lipids, and carbohydrates. He called it "nuklein:" we call it "nucleic acid," the NA in DNA and RNA.

The role of phosphate in DNA and RNA is to form, together with a pentose sugar, the "backbone" of the molecule (Fig. 2.1). It links the nucleotides, the nitrogen-based adenine (A), guanine (G), cytosine (C), and thymine (T), together to form DNA. [In RNA, uracil (U) replaces T.] These links in turn constitute the "letters" of the genetic information (Fig. 2.2).

2.3   ATP

A central molecule in the energy system of all living cells is adenosine triphosphate (ATP). It was discovered by Karl Lohmann in 1929, one of the fundamental discoveries in biochemistry that the Nobel committee mysteriously overlooked. Several other scientists, including Fritz Lipmann and Alexander Todd, later became Nobel Prize laureates for closer descriptions of its structure and functions. ATP is a coenzyme that carries out most of the intracellular energy transport. Energy is stored in

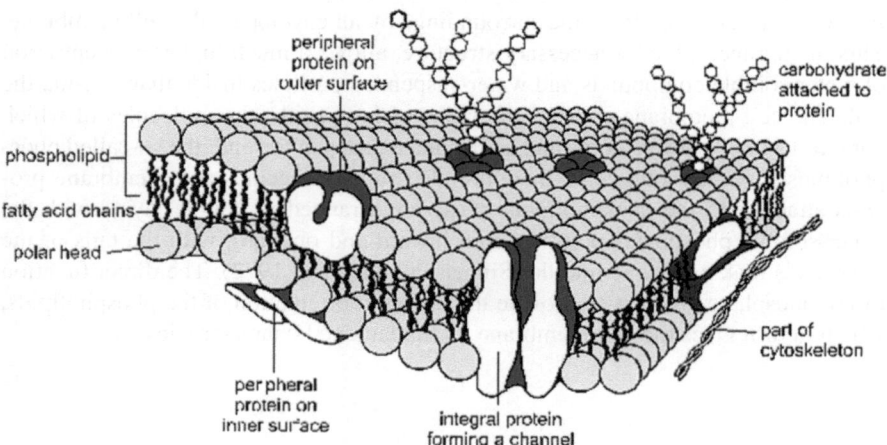

Fig. 2.1 Structure of the cell membrane (From biologymad.com)

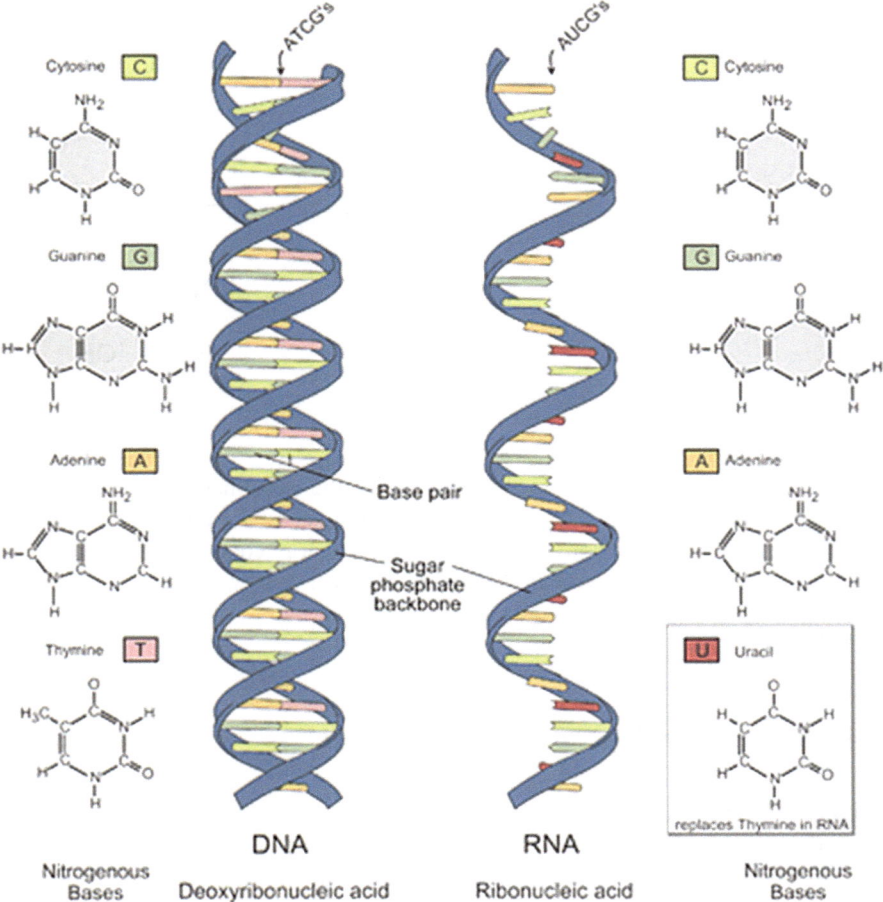

Fig. 2.2 Structures of DNA and RNA [From the database of the National Human Genome Research Institute (NHGRI)]

cells in carbohydrates such as glycogen and in fat. When energy is needed, these compounds are oxidized and energy is moved from the storage molecules to adenosine phosphate. In the most common reaction, this energy capture occurs when adenosine diphosphate (ADP) adds another phosphate group to form ATP (Boyer 1997).

Cells require energy mainly for three types of tasks: (1) to drive metabolic reactions against energy gradients; (2) to do mechanical work such as moving a muscle (mostly the co-generated heat is a loss, but sometimes it can be the main goal of the process, as when bees in a hive in winter all flex their wings); and (3) to transport substances across cell membranes. When the bond is broken through hydrolysis

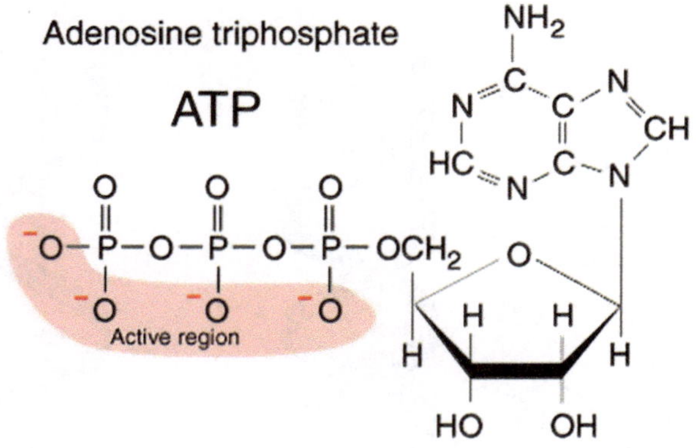

Fig. 2.3 Structure of the adenosine triphosphate (ATP) molecule

(e.g., after addition of water), ATP loses one of the phosphate molecules and is degraded to ADP. As this is an exothermic reaction, about 7.3 kcal/mol or 30.6 kJ/mol is released, which is more or less equal to the energy contained in a peanut.

Thus, the role of phosphate in ATP is to transport energy to create energy-rich bindings (Lipmann 1945) (Fig. 2.3).

2.4 Sinew, Cartilage, Bone, and Enamel

Teeth and bones, as well as cartilage and sinew, obtain their hardness and strength from the minerals they contain, different forms of the so-called biological apatite, that is, calcium hydroxyapatite with the general formula $Ca_{10}(PO_4)_6(OH)_2$. Slight differences in composition render different forms of the biological apatite in different tissues with different physical and mechanical properties.

Enamel is the hardest, but also the most brittle, with the highest concentration of apatite and thereby of phosphorus. Bone, cartilage, and sinew follow in that order. Some 85 % of the phosphorus in the body of vertebrate organisms is in the form of these apatites, which is also why bone ashes can be used as a phosphate fertilizer. Phosphorus deficiency in the vertebrate body results in fragile bones and teeth, stiff joints, bone pain, loss of appetite, fatigue, irritability, numbness, weakness, and weight loss. For growing individuals, reduced size and poor bone and tooth development can also result from lack of phosphorus. Osteoporosis in the elderly, especially women, is thought to be associated not only with insufficient intake of calcium but also of phosphorus (Heaney 2004).

References

Biologmad.com   http://www.biologymad.com/master.html?http://www.biologymad.com/cells/cells.htm

Boyer PD (1997) The ATP synthase: a splendid molecular machine. Annu Rev Biochem 66:717–749

Dahm R (2005) Friedrich Miescher and the discovery of DNA. Dev Biol 278(2):274–288

Heaney RP (2004) Phosphorus nutrition and the treatment of osteoporosis. Mayo Clin Proc 79:91–97

Lipmann F (1945) Acetylation of sulfanilamide by liver homogenates and extracts. J Biol Chem 160:173–190

Singer SJ, Nicolson GL (1972) The fluid mosaic model of the structure of cell membranes. Science 175:720–731

Chapter 3
Phosphorus in Social Life

Since about 1950, mineral oil has become the main source of external energy among all other alternatives, although it emerged in our lives only in the twentieth century. For this reason, the warning on "Peak Oil" (see Chap. 9, this volume) has become a matter of common concern, reflected in multiple energy-saving measures worldwide, creating political tensions around possession of oil reserves and causing a steady price increase trend in the oil markets.

In contrast, the vital role of phosphorus, the main source of the internal energy of every form of organic life from the times when the early forms of life appeared on our Earth, remains unnoticed. Oxygen has the same vital role in our life, and once is enough to experience suffocation to imagine "life without oxygen." But very few of us suffer from phosphorus deficiency. Thus, practically no one who is not involved with the life sciences can perceive the growing danger of "Peak Phosphorus" for all of mankind. In the meantime, as seen earlier, without phosphorus neither the creation of life nor its further development is possible.

3.1 Phosphorus and Our Body

The main functions of phosphorus in organic life are described in Chap. 2. We, as human beings, should be aware that phosphorus is the second most common chemical element among all those that comprise our body and permit its functioning. The amount of phosphorus in the human body is about 1 % of body weight, whereas in the Earth's crust the share of phosphorus is less than 0.1 %. It is not without purpose that we accumulate in our body ten times more phosphorus than exists in the surrounding environment.

About 80 % of all the phosphorus in our body is concentrated, as several forms and modifications of calcium phosphates, in the bones and teeth, providing the structural stability of the body. More than 20 different organic phosphate forms participate in dynamic processes, such as cell functioning, blood and oxygen supply

M. Butusov and A. Jernelöv, *Phosphorus: An Element that could have been called Lucifer*,   19
SpringerBriefs in Environmental Science 9, DOI 10.1007/978-1-4614-6803-5_3,
© Mikhail Butusov and Arne Jernelöv, 2013

and distribution, body growth, brain activity, and metabolism. Two prominent Russian scientists described the role of phosphorus in our life rather well: *"Phosphorus is an element of life and thinking"* (academician Alexander Fersman) and *"There is no movement without phosphorus, since the chemistry of the muscular activity is solely the chemistry of phosphorus"* (academician Vladimir Englegart).

To replenish the permanent loss of phosphorus from our body, we need to consume 1–1.2 g phosphorus per day, and those who are pregnant or engaged in intense physical activity require as much as 3 g per day. Children need extra phosphorus intake to fuel the processes of growth and development.

We very rarely experience phosphorus deficiency because we consume it in abundance in our food, although only a small part of the daily phosphorus intake is absorbed by the body. For example, phosphorus is present in some commonly consumed products as follows:

- Fish, bread, and meat: more than 250 mg/100 g
- Nuts: 400–600 mg/100 g
- Wheat bran: 1,000 mg/100 g
- Cheese: as much as 1,000 mg/100 g
- Eggs: more than 200 mg/100 g
- Apples: 10 mg/100 g
- Dates: 60 mg/100 g

The equal content of phosphorus in bread and in meat does not indicate that bread can substitute for other components of a healthy diet. First of all, the absorption rate of phosphorus from vegetable foods is slower than that from meat, fish, and eggs. Second, the healthy balance between proteins and phosphorus in our diet should be about 40:1.

A purely vegetarian diet, taken without bread and dairy products, brings a certain risk of phosphorus deficiency. Long fasting, however, does not immediately affect the main body functions, so long as the presence of phosphorus in blood remains within 3–5.5 mg %. This concentration is maintained by the release of phosphorus from different phosphates accumulated in the body (mainly in the liver), and only when the total withdrawal of phosphorus is greater than 40 % of available reserves does the phosphorus concentration in blood drop by 10 %, causing deterioration of the body functions. So, it takes only a few minutes for every one of us to experience the role of oxygen in our life, but it requires many days of fasting to recognize the equally vital value of phosphorus.

3.2 Phosphorus and Social Development

Unknowingly, many nations in different parts of the world, whose life was for many centuries strongly dependent upon agriculture, have developed ways and means of selecting the most fertile land and maintaining its fertility. Long before the discovery of phosphorus by H. Brand in 1669 (Saring 1955) and formulation of the magic triad nitrogen (N), phosphorus (P), and potassium (K) as the main components of plant nutrition by von Liebig in the 1840s, the ancient farmers knew how to distinguish fertile soil and how to maintain its fertility.

Silt, the sediments left on the soil surface after floods, contains additional nutrients for plants and increases soil productivity. This permanent fertility boost allowed the ancient countries situated in the Mesopotamia and Nile Valleys of Egypt to become rich as regional grain traders and economically more developed than their neighbors. This influence of the silt can be seen even now in some places; for example, ancient silt reserves in the Low Austria counties and Burgenland until now have guaranteed the highest quality of the local vines (Lair et al. 2009).

In those countries where river and lake floods did not cover large parts of the arable land, the primary source to maintain soil fertility was animal manure. Despite certain lack of amenities such as an unpleasant smell, possible organic contamination of surrounding water bodies, and excessive soil salinity after permanent manure fertilization, animal manure was collected in pits, preserved during winter seasons, and deployed on the arable land in accordance with their fertility: small amounts to the rich soils, larger amounts to the poor soils.

During thousands of years of experience, people learned how to rank the nutritional values of different types of manure: human, fowl, pig, cow, and horse. They also learned how to mix the manure with other ingredients to improve its positive impact on future harvests. Because ancient agriculture was mostly based on the individual and separately placed farmlands, the delivery of stored manure to the surrounding fields was not a big problem. When human rural settlement accumulated many farmhouses on a small territory surrounded by rather vast farmlands, the self-sustainability of manure-based fertilization had to be maintained by more elaborate methods of manure upkeep. This rather sophisticated culture still exists in China after many centuries of development (Olson 1987).

As the population of these human agglomerates grew, the early townships emerged and many professions were born that were not related to farming. Adequate expansion of arable lands, which was necessary to feed the growing numbers of non-farmers, became economically not viable for several reasons:

• The increasing length of transportation routes from manure storage to the fields and of the collected harvest back to the townships and
• The necessity to protect these lands from possible wrongdoers

The only solution was to learn how to intensify the productivity of available lands to foster the needs of the growing population. Until these means were found, undernourishment, famine, and epidemic events became a common menace of urban dwellers. These disasters became more serious as population density increased and elementary sanitary conditions declined. The food riots and wars that started in medieval times still continue in some continents despite spectacular developments in soil science and fertilizer production (see Chaps. 4 and 5). As one of the first manifestations of the upcoming "Peak Phosphorus," prices for basic foods for the urban population in Africa and Asia spiked. The global wheat price in 2008 was increased by 130 % compared to 2005, whereas the rice price doubled (Bush 2011). Altogether the food riots in 2008 were notable in 25 countries of our civilized world!

In some Middle East countries food riots sparked periodically from the 1980s. Despite the principally different development levels of the modern world compared to medieval times and the international mechanisms that enable us to provide the surplus of the harvest received in most developed countries to their suffering counterparts, the main reason for the food riots remains the same. Continuing urbanization cannot be sustained by the limited production capacity of farmlands.

Although the share of manure fertilization remains at the modest level of 15–17 % from the total amount of phosphorus annually brought to the farmlands, its role cannot be underestimated as the small but permanent supplier of the vital nutrients to the needs of agriculture. The new revival of the ancient skills of how to achieve self-sustainability without commercial fertilizers is demonstrated by the fast-growing interest in organic farming.

References

Bush R (2011) Food riots: poverty, power and protest. J Agrarian Change 10(1):119–129
Lair GJ, Zehetner F, Khan ZH, Gerzabek MH (2009) Phosphorus sorption–desorption in alluvial soils of a young weathering sequence at the Danube River. Geoderma 149:39–44
Olson RA (1987) The use of fertilizers and soil amendments. In: Wolman MG, Fournier FGA (eds) Land transformation in agriculture, Scope. Wiley, New York
Saring H (1955) Brand, Hennig. In: Neue Deutsche Biographie 2:515 http://www.deutsche-biographie.de/pnd135556473.html. Accessed 8 Mar 2013

Chapter 4
Silent Underground Life

Soil is sometimes called "the skin of the Earth." The health of its sensitive and vulnerable upper layer strongly depends upon global industrial development sustainability and prevailing environmental problems. As is human skin, soil is an extremely complicated substance requiring adequate understanding and maintenance. From the agricultural point of view, its role is much more sophisticated than just providing support for the plant roots and conducting nutrients through its pores from the soil nutrients toward these roots. Misunderstanding of this complexity lasted for many decades and resulted in soil destruction, desertification, and loss of its initial fertility. Further, it has resulted in vigorous attempts to boost soil fertility with overdoses of water-soluble mineral fertilizers during several decades of the twentieth century. Uncontrolled fertilization, although initially resulting in substantial increases of soil productivity, soon backfired as accelerated eutrophication, dust storms, accumulation of heavy metals in the soil, and the growing vulnerability of plants toward deceases and climatic changes. In this context, the time came to change the approach and redesign the features of fertilizers, making them an interactive component of the complex system "fertilizer–soil–roots."

4.1 Soil: The Central Part of the Phosphorus Chain

Soil is a central point of the chain *"phosphate mining–fertilizer production–crop production–food processing–human and animal consumption"* where phosphorus exerts its nutrition potential (Fig. 4.1; all data as of 1995, in Mt P/year) (Cordell 2008). In modern industrialized agriculture, P-containing fertilizer is brought into the soil, which is supposed to create substantial growth of the harvest. The harvest will further be used either for food production or for feeding livestock.

Many observations on how to grow a good harvest were accumulated by generations of farmers and gardeners during the centuries. For example, French winemakers noticed that the best grape harvest was achieved in groves cultivated on the sites

M. Butusov and A. Jernelöv, *Phosphorus: An Element that could have been called Lucifer*,   23
SpringerBriefs in Environmental Science 9, DOI 10.1007/978-1-4614-6803-5_4,
© Mikhail Butusov and Arne Jernelöv, 2013

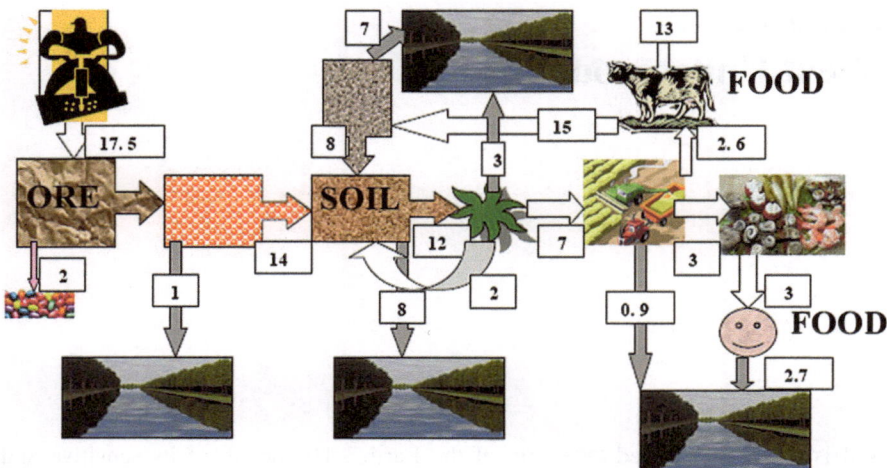

Fig. 4.1 The natural cycle of phosphorus from rock phosphate (Data in *boxes* are estimated amounts in 1995, in M tons pure P/year)

of medieval battles. But only in the nineteenth century was the crucial role of three nutrients—nitrogen (N), phosphorus (P), and potassium (K)—in plant growth and the role of mineral fertilizers intensively studied by scientists, among them Justus von Liebig, who is recognized by many as a father of agricultural chemistry and the fertilizer industry. Before this time, it was believed that plants obtained carbon from organic chemicals in the soil and acquired other essential nutrients in the form of organic compounds. Liebig showed that, in fact, plants receive their entire carbon intake in the form of carbon dioxide from the atmosphere. Furthermore, he found that to survive, plants required from the soil only water and a few minerals (such as calcium, phosphorus, nitrogen, and potassium) in the form of simple compounds. This finding meant that inorganic compounds containing only several minerals could be used to fertilize fields: organic mulch or manure was unnecessary. Following von Liebig's instructions, Muspratt and Co. in Liverpool, England, manufactured the first experimental batch of synthetic fertilizer in 1845.

4.2   Phosphorus and Mineral Fertilizers

The insights of Liebig helped to modernize the science of plant biochemistry and served as a catalyst in the development of modern agriculture and in the intense growth of the mineral fertilizer industry. The practical need for additional soil fertilization became evident in the middle of the nineteenth century. Before this time, when the world population was much smaller, farmers could obtain adequate yields by fertilizing soil with phosphorus derived from human and animal excreta.

However, the population growth stimulated higher food consumption and production, resulting in more rapid depletion of soil nutrients. To sustain their harvests, farmers had to start using increasing amounts of off-farm sources of phosphorus, such as bone meal, guano, and phosphate rock. Phosphate rock, which was cheap and plentiful, became the source that was widely preferred. Farmers also adopted new methods, such as planting high-yielding crop varieties and then applying mineral nutrients, notably nitrogen, phosphorus, and potassium (N, P, K), and other inputs such as pesticides. The so-called "green revolution" commenced.

However, in the middle of the twentieth century it became apparent that the findings of von Liebig had suffered inaccurate interpretation. In a simple way, until that time the plant fertilization process (with emphasis on the transformation of phosphorus) was understood as follows:

- After dissolution in soil water (water is a vital prerequisite of any successful fertilization), the P-based fertilizer generates two types of anions of phosphoric acid: $H_2PO_4^-$ and HPO_4^{2-};
- These anions are transported by the water flow toward the sown seeds or toward the plant roots
- Arriving at the vicinity of seeds and roots, the P anions are absorbed by them and render their productive impact on plant growth and development

Excessive consumption of mineral fertilizers accelerated in the last half of the twentieth century with the Green Revolution and created serious environmental problems, as described further in Chaps. 5 and 7.

4.3   What Really Occurs to Phosphates in Soil?

The oversimplified picture of how fertilizers function in the soil was too nice and peaceful to be true. The underground life is full of conflicts and silent battles; it is not as straightforward as people believed for many decades.

A convincing test of what happens to the phosphorus dissolved in soil water was made on a pasture field by the scientists of the Lincoln University in New Zealand (McLaren and Cameron 1996). They made several measurements in the topmost productive layer of the pasture soil (about 30 cm thick).

First, they measured a total amount of phosphorus contained in the top layer of unfertilized (virgin) soil and found out that the average phosphorus (P) concentration is *650 mg P kg⁻¹* soil. This amount was found to be equivalent to about *1,000 kg P/ha* (hectare) placed in the top (root-containing) layer of soil.

Then, in due time the grass was harvested from this soil. One hectare of pasture produced about 14 t plant material (dry matter, including roots) containing 0.35 % phosphorus. Thus, this harvest absorbed only 49 kg phosphorus per hectare from the soil in 1 year. The amount of phosphorus extracted by the harvest was *more than 20 times less* than the total amount of phosphorus stored in the root-containing layer!

Despite this, the majority of unfertilized soils in this region were found to be originally deficient in phosphorus for maximum pasture growth. Yields on these soils were typically 25 % of the production that could have been obtained when adequate amounts of phosphate fertilizers were used. The main question then arose: where was the rest of the phosphorus?

In further experiments, enough phosphate fertilizer was applied to produce maximum plant growth. Again, the amount of phosphorus extracted by the harvest remained much lower than the amount of applied phosphorus. Additionally, production declined in each successive season if the phosphate fertilizer was not reapplied regularly.

Next, the soil water in the same pastures was tested. It was found that the water typically contained only about 0.05 mg l^{-1} inorganic phosphate in solution, which is equivalent to an unbelievably small amount of about *15 g phosphorus in solution in 1 ha* of land!

Comparing these figures:

- Total amount of phosphorus contained in the topsoil (about or more than 1,000 kg/ha)
- The amount of phosphorus extracted by the harvest (about 50 kg/ha)
- The amount of phosphorus contained in the soil water (about 15 g/ha)

One can assume that the straightforward model describing soil as just a neutral medium holding the plant roots and providing transportation of the P-carrying soil water to these roots is incorrect. Only fractions of a percent of the total P content in the soil are present in the soil water at any moment in time.

It is hard to say whether these or similar data were also timely received by the relevant scientists of the leading research centers of Europe and the United States, and if yes, why did not they initiate open discussions concerning one of the most crucial problems in agronomy: the interaction of mineral fertilizers with the soil and plants?

Apart from purely scientific interest, these controversies were crucial from the economic and environmental aspects. For example, the production and use of water-soluble fertilizers should have been doubted. Indeed, if only a slight amount of the applied phosphorus exists in the soil in stable water solution, what can be the benefit of using commercial fertilizers, most of which are immediately water soluble (Table 4.1)?

4.4   Should the Effective Fertilizer Be Water Soluble?

For example, if the main role of the P fertilizer is not only to be a source of an immediate phosphorus release to the soil water, if the soil itself plays an active role in transformation and transportation of relevant nutrients to the plant, why then was it required that phosphoric fertilizers should be water soluble to be effective?

Table 4.1 Percentages of water-soluble phosphates in several common fertilizers in the United States

Type of fertilizer	Chemical formula	N-P-K equivalent (%)	Solubility (%)
Ammonium polyphosphate	$NH_4H_2PO_4 + (NH_4)HP_2O_7$		
Liquid		10-34-0	100
Dry		15-62-0	100
Diammonium phosphate	$(NH_4)2HPO_4$	18-46-0	≥ 95
Monoammonium phosphate	$(NH_4)H_2PO_4$	11-48-0	92
Single superphosphate	$Ca(H_2PO_4)_2 + CaSO_4$	0-20-0	85
Triple superphosphate	$Ca(H_2PO_4)_2$	0-46-0	87
Rock phosphate	$3Ca_3(PO_4)_2{}^a \cdot CaF_2$	0-32-0	≤ 1

Source: Schulte and Kelling (1996)
[a]Water-soluble data are a percent of the total P_2O_5

Even now, when it has become evident that excessive leaching of the colloidal and soluble forms of phosphorus into surrounding aquatic reservoirs causes serious environmental problems, such as eutrophication (see Chap. 7), most of the produced fertilizers are water soluble, despite the fact that only a tiny part of their P content can exist in a water solution (Table 4.1).

The first doubts about the necessity of water solubility for fertilizers appeared quite some time ago. Already in 1953 L.T. Kurtz (1953) suggested that the soil might play a much more active role than was understood at that time:

> Contrary to the apparent belief…, more recent evidence indicates that the reactions of phosphate with soils are not entirely irreversible…

This statement meant that the soil could temporarily capture phosphorus dissolved in the soil water and, further on, release it in certain conditions for effective plant nutrition.

Data presented further in Chap. 5 show that most of the industrialized countries became, as of the beginning of the twentieth century, strongly dependent upon several global producers of "standard" (mostly water-soluble) phosphoric fertilizers. However, there were not sufficient incentives for the agricultural community to reconsider the traditional picture of how the fertilizers interact with the soil and plants.

For a long time the same old pathway was followed:

> The more phosphorus is washed out from the productive soil layer, the greater the amounts of fertilizers are necessary next year and the higher remain the profits of producers, wholesalers, and other major players in the fertilizer market.

It may explain also why much more interest developed in making phosphoric fertilizers more effective and, by that means, minimizing their consumption, in countries not possessing indigenous phosphate resources and depending upon the import of costly fertilizers from remote suppliers. Such countries included Australia, New Zealand, and several countries of Southeast Asia (see, for example, Syers et al. 2008)

As a result, the new soil concept was elaborated, has undergone many tests, and is now gradually becoming accepted in most countries.

Phosphorus in the soil may exist in four different pools, in which its stability can differ greatly:

- *Soil solution*, from which P is immediately available for the plants
- In the *first, readily available*, state where P is slightly bound to the soil particles
- In the *second, less available*, state where P is rather strongly bound to the soil particles
- In the *third* state, where P is practically irreversibly bound to the soil

Between these pools, there exists a permanent exchange of phosphorus substances caused, most possibly, by the following chemical reactions and physical processes.

1. After a granule of P fertilizer (for example, superphosphate) is inserted in the soil, it results in high phosphate concentrations existing in the soil water around this granule. This solution has rather low pH and may interact with the nearest surrounding soil particles that normally contain substantial amounts of hydrous iron and aluminium oxides and aluminosilicates. As a result, temporary formation of iron and aluminium phosphates occurs. The remaining monocalcium phosphate, MCP ($CaHPO_4$), from superphosphate will be mainly converted to dicalcium phosphate, DCP ($Ca(H_2PO_4)_2$). As a result of these two reactions, the immediate effect of the fertilizer on the concentration of phosphate in the soil solution will be controlled by the solubility of the newly formed phosphates. However, this is only a temporary phase.

2. The soil solution containing these phosphates spreads from the fertilizer granule outward. Then, these inorganic phosphates are rapidly adsorbed from the soil solution on the surface of the soil particles. The amount of the adsorbed phosphates, the speed of this reaction, and, conversely, their resulting concentration in the soil water—these parameters depend on the nature of the adsorbing surfaces. Careful studies made on the surface properties of the soil particles surrounding freshly applied fertilizer granules demonstrated an increased presence of surface complexes containing P. Main agents of P adsorption are several inorganic substances, such as those containing Al and Fe, as well as organic inclusions such as humus. Those phosphates that are not adsorbed and still remain in the soil water have increasing chances to be, later on, leached out from the upper (about 30 cm thick) most productive layer of soil. Not only do these phosphates become useless from the nutritional point of view, they join the streams of soluble nutrients dispatched to nearby rivers and lakes and there contaminate aquatic flora and fauna (eutrophication).

3. Because there are several mechanisms of P fixation on the surface of soil particles, these fixation points can be described as multiple shallow traps where phosphates are caught. This process is reversible. As soon as the phosphate concentration in the soil water drops (as a result of either plant consumption or leaching), phosphate ions can be released and replenish the P deficiency in the

soil water. Thus, the second pool of phosphates, the adsorbed phosphates, are in fact a source of readily available nutrients that can be provided "on demand" to the soil roots. If this is true, the whole traditional concept of P fertilizers should be revised. The pools of the most productive phosphates in the soil are these areas of the readily available phosphates on the surface of soil particles, not the fertilizer granules themselves! New understanding suggests also that using the fertilizers that dissolve more slowly than conventional superphosphates and ammonium phosphates will also decrease hazardous leaching of P during the entire year after fertilization. Truly effective fertilizers should offer a much more gradual increase of phosphate concentrations in the soil water than do the highly soluble phosphate fertilizers. If the rate of release of phosphates from the fertilizer is similar to the demand of the plant for the phosphates, the plant will be able to compete more effectively with leaching, P absorption, and its immobilization in the soil (see following).

4. A second mechanism following and competing with surface adsorption is caused by gradual penetration (diffusion) of P-containing substances to the inner volumes of the soil particles. A similar process takes place in the soil aggregates, or crumbs. Soil aggregates are clusters of soil components held together by a number of mechanisms that include both organic and inorganic cements and have charged surfaces. Aggregates contain a fine network of pore spaces whereas the aggregates themselves are separated by a coarser system of pores responsible for water and air movement within the soil. When a phosphate-containing solution passes through the soil, phosphate is first removed from the solution by adsorption on soil particles located at the surface of soil aggregates. Part of this adsorbed phosphate then subsequently diffuses into the aggregates. Absorbed and caught inside the tiny pores, phosphates represent a third pool of phosphates in the soil that has much lower availability for the roots. Only when the soil is for a long time undernourished this pool can release a part of the trapped phosphates back to the soil water. Another factor resulting in the partial release of absorbed phosphorus is acidity changes in the soil water (Koutsoukos and Valsami-Jones 2004), as illustrated by Fig. 4.2.

5. Most of the absorbed phosphates gradually undergo crystallization, occlusion, bonding with organic materials, and other chemical processes, making these phosphates practically unavailable for nutritional purposes. Indeed, measurements of the total amount of phosphorus contained even in poorly productive soil show that this amount may exceed the needs of plants or pasture for several rather good harvests. However, most of the phosphorus reserves in the fourth pool are made dormant and essentially unavailable.

Many factors influence P dynamics, that is, the rate of binding and release of P substances to and from each pool and their transfer to another pool. Among these factors, the primary ones are as follows.

Basic properties of the soil before fertilization:

1. Soil acidity level (pH)
2. Amount of Ca ions in the soil

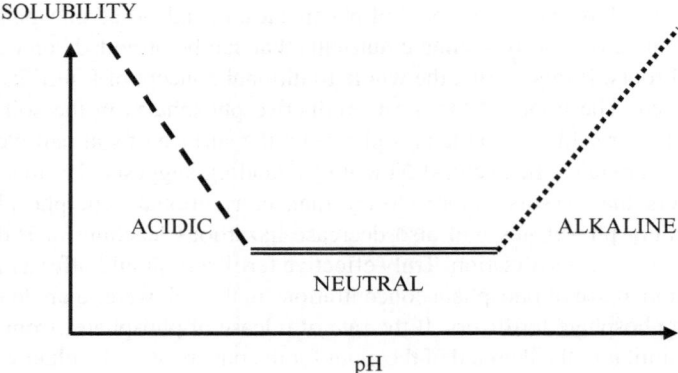

Fig. 4.2 Absorbed phosphorus dissolution rate as a function of the soil pH

3. General soil fertility level (before application of the phosphate)
4. Presence of organic matter in the soil
5. Average size of the phosphate particles

  Properties of fertilizers and external conditions:

6. Placement of phosphates (on the surface or subsurface), and time and rate of application
7. Climatic conditions (such as seasonal rate of rainfall and snowmelt)

  Most of these rules have been confirmed during field experiments. In short, they can be explained in the following way:

1. Soil acidity influence (illustrated by Fig. 4.2). The general mechanism of dissolution of slow-release fertilizers in the soil is described by the following reaction:

$$Ca_{10}(PO_4)6A2 + 12\ H_2O \Rightarrow 10\ Ca^{2+} + 6H_2PO_{4^-} + 2A^- + 12OH^-$$

  where A symbolizes certain anions present in soil, such as Cl^-, F^-, or OH^-. This equation means that dissolution of phosphates results in release of the hydroxyl ions into the soil water. The faster these ions are captured by the soil particles (which happens faster when $pH \leq 5.5$), the more the equilibrium in this reaction is moved to the right side, and the phosphates continue to dissolve.
2. Presence of Ca ions. Ca ions are a part of the soil property called "cation-exchange capacity." Primarily, additional Ca^{2+} cations released during dissolution of phosphates are removed from fertilizer particles by soil adsorption. Then, the probability that Ca^{2+} ions rejoin the phosphate ions again and that the initial reaction is reversed becomes lower. The residence time of phosphate anions, $6H_2PO_4^-$, in the soil water is thus prolonged. Another way to block Ca^{2+} ions is

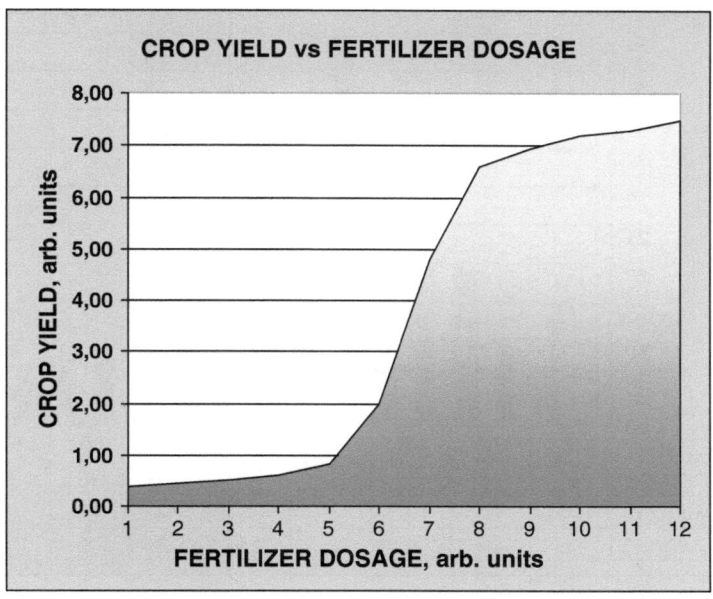

Fig. 4.3 Dependence of crop yield upon amount of applied phosphates

to apply certain polymers in the soil together with phosphates. A typical example is the product AVAIL of the company SFP (www.chooseavail.com), a water-soluble polymer with double action. It partly encapsulates the phosphate parti-cles in the soil by the water-soluble "shield" from the blocking activity of Ca^{2+} cations and partly surrounds these cations and reduces their negative influence on the concentration of phosphate ions in the soil water.

3. The influence of soil fertility level, its potential productivity before application of phosphates, can be explained in the following way (Fig. 4.3).

   If soil fertility before application of phosphates is low (see lower zone on Fig. 4.3), there is a high concentration of unsaturated Ca "traps" on the surface of the soil particles. Those phosphate ions that are slowly released from the applied grains are effectively immobilized by these traps. A medium fertility level means that soil conditions are in the medium zone: most of the Ca "traps" are blocked by the indigenous phosphate ions, and additional P-carrying ions released to the soil water by the applied phosphate have more chances to travel toward the plant roots. In the upper part of the curve where the soil itself is fertile enough, is clearly also rather insensitive to the application of additional phosphates.

4. Soil organic content. Application of external phosphates in the soil with high organic content is rather effective for two reasons:

   • Certain organic substances, such as legumes and humates, are known for their ability to discharge specific organic acids, by that means further reducing soil pH and therefore increasing phosphate efficiency (see Fig. 4.2)

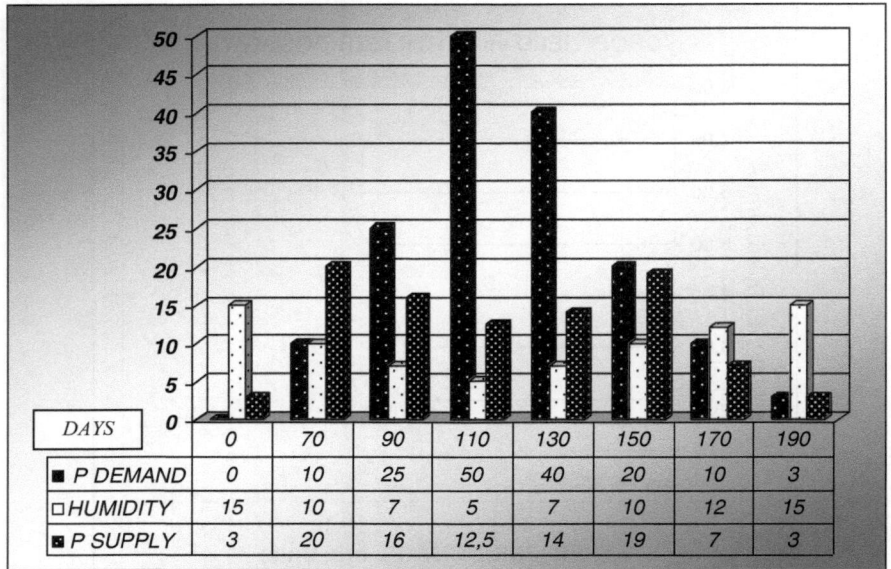

Fig. 4.4 Soil humidity (*white bars*), P demand (*black bars*), and P supply from water-soluble fertilizer (*dotted bars*) during 1–170 days after sowing

- Ca²⁺ ions residing on the surface of the soil particles produce rather stable complexes with much organic matter. Therefore they cannot participate in blocking the phosphate ions released into the soil water.

  Unpublished pot tests made by one of the authors with slow-release P fertilizer blended with small amounts (0.5–1.0 %) of synthetic humates produced from pulp and paper industry wastes showed undoubtedly the increase of soil productivity.
5. Fineness of the phosphate and soil particles. It is clear that the smaller the phosphate particles, the higher is their interaction surface with the soil water, and the more phosphate ions are released to the soil water. However, placement of fine-milled fertilizers in real field conditions may result in dusting and essential loss of the product. Also, excessive rainfall may wash out the fine particles from the productive soil layer. The soil particle size factor defines its cation-exchange capacity. For example, sandy soils have low capacity, and field tests have confirmed that on these soils the efficiency of slow-release fertilizers is decreased.

  Parameters independent from the soil properties that can also influence the agronomic efficiency of the natural phosphates are the following.
6. Soil management practice

  - Amount of the soil water during the season. Soil humidity in normal European conditions is the highest after the snowmelt in spring (days 0–70 on Fig. 4.4) and during the autumn rains (days 150–190). At the same time, most of the plants experience the largest demand in P nutrients between 90 and 150 days after sowing, where the amount of soil water drops. If water-soluble fertilizer is applied, its enhanced concentrations in the beginning and at the end of the season are unneeded and lead to eutrophication.

- Placement of fertilizer granules. Because of the adverse effects of application of fine-powdered material, most fertilizers nowadays are used in the granular form. In view of the described P-transport picture in the soil, it becomes essential how these granules are placed in relationship to the plant roots (see following). To achieve higher efficiency, granular fertilizer should be broadcast (i.e., applied uniformly over the field) and incorporated into the soil surface at the depth 100–150 mm.
- Rate of application. If the soil fertility status is medium to high (medium zone, Fig. 4.3), then the rate is defined on the basis of the annual plant P-consumption, so that the fertility level is kept permanent. In other words, it depends upon the type and amount of plants on the fertilized area, thus requiring agronomic assessment. If the soil fertility is low (first zone on Fig. 4.3), then the farmer should first increase the fertility level, either by initially applying liquid mineral fertilizers or by applying milled rock phosphates in abundance (500–100 kg/ha).
- Timing of application. For acidic soils (pH$\leq$5.5) with high P-retention capacity, it is recommended to incorporate P material close to the time of planting. If P-retention capacity is weak (pH$\geq$5.5), then it is better to apply the phosphate ahead of planting time.

This comprehensive picture opens a path for new understanding of how P fertilizers interact with the soil. The following conditions should be observed to maintain the soil productivity:

- Adjust the type and amounts of fertilizers actually necessary for a given soil, climate, and culture
- Avoid or alleviate eutrophication caused by P-leaching
- Avoid soil destruction and the tendency of many plants to take up more phosphate than they need, termed 'luxury consumption." Indeed, luxury consumption of plants reminds us of the negative consequences of human obesity. Plants need to be kept under certain limited stress conditions. They need to be "trained" to develop their own strong root structure and fight for the nutrients, by that means competing with the effects causing the irreversible intake of nutrients by the processes occurring inside the soil particles.

4.5 Plant Roots in Competition for the Nutrients in Soil

Plant roots have developed several tools to acquire the necessary nutrients from soil, thus adjusting their demands with the adsorption/absorption processes in the soil. The main mechanism is the highly developed root system. A root system of field plants can consist mostly of two types (Fig. 4.5):

FIBROUS ROOT   TAPROOT

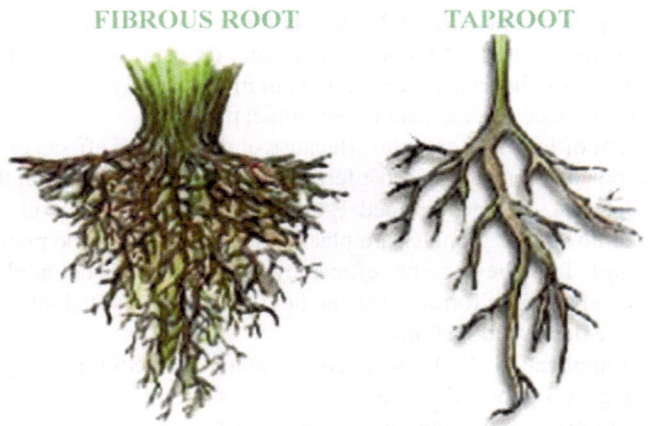

Fig. 4.5 Fibrous root (*left*) and taproot (*right*) systems

- Fibrous root systems consist of many fine hairlike roots that form a thick mat below the surface. These root systems are very effective at water and mineral absorption as well as plant stabilization. Fibrous-rooted plants are very effective at preventing soil erosion and promoting slope stability because of the formation of fibrous root systems near the soil surface. Examples include native grasses, sedges, and rushes, as well as most colonizing wildflowers.
- The taproot system contains one large, primary vertical root that spawns many smaller horizontal root structures. By penetrating deep in the soil, taproots provide stability and store nutrients. Plants with large taproots should not be used when the soil is very shallow. Examples include coneflowers, milkweeds, indigos, and certain trees and shrubs.

The importance of the root system for any plant is defined by the following features:

- Fixing the plant in the soil
- Extraction and transmission of the soil water and relevant nutrients to the surface part of the plant
- Accumulation of nutrients in the root system as well
- Symbiotic interaction with fungi and other microorganisms domesticated in the soil

The last feature has a special importance because most of the field plants develop on their roots large colonies of fungi, the mycorrhizae, allowing them to establish mutually beneficial relationships. These specialized fungi colonize plant roots and extend far into the soil. Mycorrhizal fungal filaments in the soil are truly extensions of root systems and are more effective in nutrient and water absorption than are the roots themselves. More than 90 % of plant species in natural areas form a symbiotic relationship with the beneficial mycorrhizal fungi.

In a simple way, the symbiotic coexistence of plants and fungi works as follows: when the plant does not have enough nutrition from the soil, mycorrhizal fungi also do not receive enough food from their "masters"—the plant roots. In this case the fungi release powerful enzymes into the soil that dissolve hard-to-capture nutrients, such as organic and inorganic phosphorus, iron, and other tightly bound soil nutrients. Referring to Fig. 4.2, this biochemical activity of enzymes can be described as causing a decrease of pH in the vicinity of roots, by that accelerating elution of the absorbed and otherwise bound phosphorus to the soil water and, respectively, accelerating intake by the plant roots. When the plant no longer senses phosphorus deficiency, the release of enzymes by fungi stops, pH level recovers, and no more phosphorus is released to the soil water. These "interactive" relationships among roots, soil, and phosphates indicate a new concept of effective fertilizers.

References

Cordell D (2008) The story of phosphorus: missing global governance of a critical resource. In: Proceedings of the SENSE Earth Systems Governance, Amsterdam, 24–31 Aug 2008

Koutsoukos FG, Valsami-Jones E (2004) Principles of phosphate dissolution and precipitation. In: Valsami-Jones E (ed) Phosphorus in environmental technologies. IWA, London

Kurtz LT (1953) Phosphorus in acid and neutral soils. In: Pierre WH, Norman AG (eds) Phosphorus in crop nutrition, vol IV, American Society of Agronomy. Academic Press, New York

McLaren RG, Cameron KC (1996) Soil science: sustainable production and environmental protection, 2nd edn. Oxford University Press, USA

Schulte EE, Kelling KA (1996) Publication A2520: "Soil and applied phosphorus". University of Wisconsin, http://www.soils.wisc.edu/extension/pubs/A2520. Accessed 25 June 2012

Syers JK, Johnston AE, Curtin D (2008) Efficiency of soil and fertilizer phosphorus use. FAO, Rome

Chapter 5
Fertilizers: 100 Years of Supremacy

Urbanization gradually started in Europe in medieval times, that is, between A.D. 500 and 1500. Historical need and the necessity of urbanization were created by the multiple advantages of living together in large cities, as compared to being dispersed across vast territories. However, this migration was followed by several negative factors, the influence of which cannot be underestimated, such as the growth of poverty and inequality and the deterioration of sanitary conditions and the health of populations. The human nutrition chain also was perturbed: people could no longer live by means of their fields and livestock, and cities needed the development of more extensive farming nearby to supply city dwellers in increasing numbers. In the beginning, natural soil fertility in the surrounding rural areas allowed achieving better harvests, but with the continuous increase in food consumption in urban settlements and the shrinking of farmers' land around the cities, malnutrition of large parts of the urban population became a factor of social stress and economic danger. The discovery of fertilizers was a breakthrough that allowed further industrialization and urbanization to continue until the end of the twentieth century.

5.1 Rock Phosphate: The Only Remaining Natural Phosphorus Resource

From Chap. 4 it seems inevitable that external nutrients, mostly P, N, and K, should be put in the soil in huge quantities if we want to supply the ever-growing global population with an adequate food supply. It means that mining natural phosphate reserves and processing the phosphate rock to fertilizers will remain a growing industrial activity despite the several huge environmental problems this activity creates, such as these:

1. Depletion of limited and, in some continents, even exhausted, natural deposits of phosphate rock

M. Butusov and A. Jernelöv, *Phosphorus: An Element that could have been called Lucifer*,   37
SpringerBriefs in Environmental Science 9, DOI 10.1007/978-1-4614-6803-5_5,
© Mikhail Butusov and Arne Jernelöv, 2013

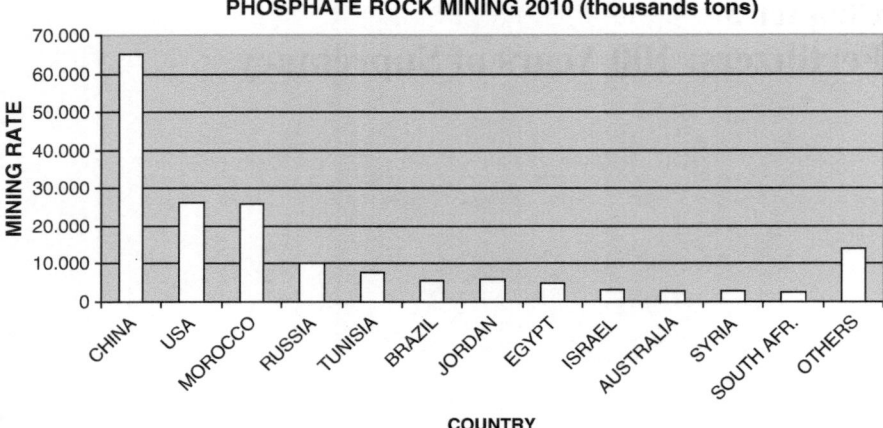

Fig. 5.1 Global production, breakdown by main countries: producers, estimates (From Jasinski 2012)

2. Destruction of vast territories surrounding mining sites and fertilizer production plants, including enormous deposits of tailings and polluted water
3. Large energy consumption related to the production and transportation of fertilizers
4. Gradual increase of the presence of some heavy metals, such as cadmium, and of radioactivity, because both factors are indigenous for sedimentary phosphates
5. Leaching of a substantial part of the phosphoric substances from agricultural areas cultivated with the standard water-soluble fertilizers and polluting of surrounding bodies of water (eutrophication)
6. Damaging soil structure and plant endurance as a result of overfertilization

Because pure phosphorus is an extremely active chemical element, its global deposits are mainly represented by phosphorites, ore structures composed of apatites, which are different salts of the phosphoric acid $H_3(PO_4)_2$, mostly defined as orthophosphates with the general chemical formula $Ca_5X(PO_4)_3$ (where X is a fluorine, or, less often, a chlorine or a hydroxide group OH) and accompanied by or embedded with shales, cherts, limestone, dolomites, and sometimes sandstone. Reviewing the chain from the natural phosphate rock to the final product, one sees that this chain contains large flaws. First, known reserves of P rock are very unevenly scattered across the globe and are mined at different rates (Fig. 5.1).

The second problem is the evident limitations in size of natural reserves and production capacities. The global P-rock mining rate achieved 176 million tons in 2010. In first view of the estimated global P resources of 65,000 million tons, this amount seems rather modest and there is no apparent problem with ore exhaustion: known resources can be mined for the next 350 years! This argument of those who call themselves "anti-alarmists" and wish to continue "business as usual" is, however, at fault.

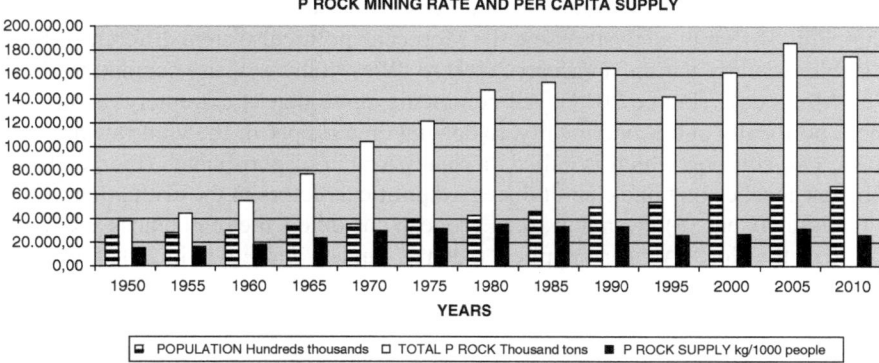

Fig. 5.2 Phosphate rock production rate (in 1,000 t/year), global population (in hundred thousands people), and P rock supply per capita (in kg/1,000 people) during the past 60 years

Correct interpretation of resources attributes these 65,000 million tons to the amount of P rock *of any grade that may be produced at some time in the future, disregarding production costs.* At the same time, the reserves, that is, the amount of P rock that can be economically produced in the form of P concentrate and is appropriate for making suitable products, are estimated as only 16,000 million tons. One can still say, "OK, we have enough reserves for the next 60 years" in the hope that during the twenty-first century some new technologies will be developed allowing us to mine more and more cheaply. It is hard to believe. Even more peculiar ideas postulate that the estimated concentrations of P in the crust of Mars are 20 times greater than in the Earth crust, so that P rock can be mined there (Reddy 2004). The question is whether Mars is going to come closer within the next 100 years?

If we concentrate on our earthly problems, we already see the first signs of the increasing redundancy of available phosphorus (Fig. 5.2).

The permanent growth of the population (from 2.5 billion by the end of 1950 to 6.5 billion by the end of 2010), a stable but rather small supply of manure as an additional source of phosphorus (about 2.5 M tons in pure P per year), and the reduction of fertile terrain used for food production require a disproportionate increase of P-rock mining rate. Figure 5.2 shows that as from 1990 this was not the case. Therefore, a warning about "peak phosphorus" emerged (see Chap. 9).

As a consequence of this warning, even if the current mining rates continue, the existing reserves will be exhausted in the United States within 54 years and in China within 57 years. In the United States this concern was taken seriously: national mining rates decreased by 32 % from 1999 to 2009, and the country restarted extensive import of phosphates. China had to double their mining rates during the same time and stop exporting raw phosphates for internal reasons: its growing population and changes in the national diet from traditional soybeans to poultry.

In the meantime, most of the countries in the world, and European countries among them, had to rely on the few remaining suppliers of phosphates, first of them being Morocco (Fig. 5.1)

How reliable is this situation of the complete dependence of Europe on one supplier? Different analysts assess the Moroccan political system differently, from "stable, friendly nation" (Vaccari 2009) to "West Sahara is an occupied territory (by Morocco)" (Pierce 2011). All comments agree that is extremely dangerous, both politically and economically, to depend on a supply of the vital raw material from one country, which is called "a Saudi Arabia of phosphate" (Pierce 2011). History teaches that one cannot disregard political factors in the evaluation of this vital issue (Chap. 8): even now, for European customers the remoteness and instability of the North African suppliers (such as Tunisia and Egypt in 2012) is a serious risk factor and always raises the question of added costs (see following).

5.2   Production of P Fertilizers

Production of mineral fertilizers started in the nineteenth century, and the first method of acidification of P rock was patented by John Bennet Lawes in 1842 when a patent was been granted to him for the development of superphosphate: bone meal, or calcium phosphate, treated with sulfuric acid. Superphosphate was the first artificial fertilizer, which his factory produced for the next 30 years. The idea of making these two raw materials more readily soluble by acidification stemmed from Lawes' experiments in his estate at Harpenden, and from 1843 at the Rothamsted Experimental Station, where the first agricultural research station in the world was established. Experiments were conducted on different fertilizers. They tested crops that were previously normally grown in rotation, but they started using the practice of growing the same crop year after year on the same plot using a variety of manures and fertilizers. More than 100 papers were published by Lawes and his lifelong partner H. Gilbert.

Combining his research with successful business, Lawes established the first plant producing mineral fertilizers on his methods, and was producing, by the 1870s, about 40,000 t superphosphate per year. Rock phosphate recommended itself as a more reliable and cheap raw material than bone and was mostly used for production.

Production of P-based mineral fertilizers for the next 160 years was based on the same technology. Further modifications in the production range, such as P-N or P-N-K fertilizers, required more sophisticated processing, but the basic technologies for production of their P component remained as before (Fig. 5.1). Current requirements for the parameters of rock phosphate are mostly based on the economical viability of mining and processing:

- Concentration of P_2O_5 between 23 % and 37 % (higher concentrations are practically not found)
- Low free carbonate content to avoid excessive consumption of sulfuric acid (see below)
- Low Fe_2O_3, Al_2O_3, and MgO content to avoid formation of intermediary products during the production cycle
- Low Cl^- content to avoid corrosion of the equipment

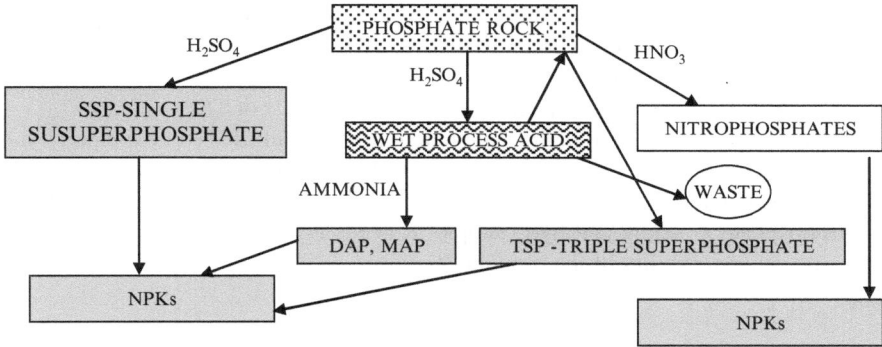

Fig. 5.3 Main steps of production of P-based fertilizers

As we see, significant environmental concerns such as the presence of toxic impurities, first of all, heavy metals and radioactive elements, are not considered.

Mining and processing the phosphate ore has the following main steps:

- Most (>80 %) of P rock used in fertilizer production is sedimentary in origin: that is, these rocks were formed in the continental shelf marine environment and are thus taken from present or former continental margins. The rest is mined from igneous reserves that were formed mostly in shield areas and rift zones.
- Sedimentary rock is mined in open pits. Before mining, the pit must be opened, and 5 to 10 m of the surface sands and clay is removed and put aside.
- Then, the ore is mined and pre-processed (Fig. 5.3 and Fig. 5.4).
- Treating the fine-milled P rock with sulfuric acid at a concentration above 57 % achieves a mixture of Ca monophosphate and anhydride (dry Ca sulfate), whereas the auxiliary by-product, gaseous HF, is removed from the reaction zone:

$$\left[Ca_3\left(PO_4\right)_2 \right]_3 \cdot CaF_2 + 7H_2SO_4 + 3H_2O = 3Ca\left(H_2PO4\right)_2 \cdot H_2O + 7CaSO_4 + 2HF.$$

The product is called single superphosphate, SSP. Because the initial high-quality P rock contains about 26–29 % P_2O_5, SSP is diluted by gypsum to 18–19 % P_2O_5. This is a main drawback of this process; however, it would be untrue to say that presence of gypsum is always a drawback of superphosphate. For the podzol and sandy soils, which are devoid of Ca and S, and for such cultures that require additional sulfur (legumes), SSP is rather effective. In most cases, however, the presence of gypsum is a clear disadvantage.

- If the phosphoric acid, also produced by acidification of P rock, is dehydrated and again mixed with the P rock, the concentration of P_2O_5 in the final product can be increased to 40 % and so-called triple superphosphate (TSP, 0-40-0) is produced. In this case most of the gypsum ends up as waste (so-called phosphogypsum).
- Complex fertilizers as NPK (with different percentages of the three components) and two water-soluble NP fertilizers, MAP (monoammonium phosphate 11-52-0) and DAP (diammonium phosphate 18-46-0), are achieved with the addition of nitric acid and ammonium into the production cycle.

The general comment on all these technologies is that they require huge quantities of chemicals in amounts similar to that of the initial P rock, and, in many cases, a large consumption of electric energy. In case of TSP, about 4 t P rock, 1.3 t pure quartz, and 130–150 kW electric energy are needed for 1 t of TSP.

If the stoichiometric ratio in the main reactions is not fulfilled (for example, if the raw P rock is richer or poorer than the required level), the final product is insufficient and needs after-processing. If a free phosphoric acid remains, the produced superphosphate is acidic and hydroscopic: it needs neutralization. In the opposite case, the water solubility of the product decreases.

5.3 Additional Technologies Used in Production of P Fertilizers

Apart from the basic chemical processes, production of P fertilizers requires many additional steps such as follows:

- Opening the pit by removing 5–10 m of the overburden (sands, clays, and their mixtures)
- Ore extraction using mining techniques such as draglines, bucket wheel excavators, front-end loaders, and removal paddlewheels
- Beneficiations, such as calcination (to remove organic matter) and screening (to remove oversize material)
- Washing by hydrocyclones to remove clay inclusions
- Flotation to increase the pure phosphate content by removal of silica-containing fractions

Most of these operations take place either on site or at the nearest site to avoid transportation of huge masses of the ballast.

Because most P mines are remote from the end-users, a substantial part of the supplementary technologies also includes logistics (shipping by ships or trucks), final beneficiation (such as Cd removal from the Morocco subproducts in Europe),and production of complex fertilizers. More and more often production of phosphoric acid is also outsourced to the on-site operations. Disregarding environmental hazards, this decision is driven mostly by profit maximization by the initial producers.

5.4 Environmental Consequences of P Fertilizer Production and Application

Multiple environmental hazards are typical for the current large-scale production of mineral fertilizers. Some of them are caused by the producers' ignorance, but mostly the main driver is profit maximization under conditions of hard competition.

Scrolling along the whole production cycle, it is possible to mark most of these hazards.

Fig. 5.4 Open pit phosphate mining in Togo (From UNEP 2011)

5.4.1 Mining Sites

Large heaps consisting of surface overburden and refuse ore surround the open pits (Fig. 5.4).

5.4.2 Processing Sites

One of the polluting factors is the wastewater related to washing the slime generated during the clay screening. The amount of slime depends upon the characteristics of the natural ore. Slime is handled in the same way as during the past 40 years: stacked, dumped into the sea, or placed in settling ponds. If the mining or processing sites neighbor open waters, contamination of natural reservoirs with the solid, colloidal, and soluble waste causes eutrophication and damage to fisheries and recreational areas, etc. (Fig. 5.5). The state-owned company OCP is the main owner and operator of the largest phosphate mines and processing plants in Morocco. Not much was said about the environment protection measures of OCP during the first International Symposium on Innovation and Technology in the Phosphate Industry, held in Marrakech, 9–13 May 2011. It is known that the discharges from at least some processing plants, situated close to the coastline, are channeled directly to the sea, but foreign participants were not granted permission to visit the relevant Moroccan facilities. One can only guess why. Phosphate mine tailings normally contain two types of waste: fine-grained muddy tailings and

Fig. 5.5 Sea pollution at coastal area of Togo caused by phosphorus rock washing plant (From UNEP 2011)

coarse-grained waste. There is a possibility of even greater negative impacts on the environment of the Atlantic because OCP performs not only mechanical but also chemical processing of the phosphate ore (production of the phosphoric acid) directly at the coastal area.

A study made at the coastal zone of Togo where annually 3.5 M tons of phosphate rock are mechanically processed (wet sieved) reports (Gnandi et al. 2006) that the largest part of the waste streams, accounting for 40 % of the rejected raw material in the amount of ~2.5 M tons annually, is directly dumped in the coastal waters of Togo as a slurry that is highly enriched with Cd, Cr, Cu, Ni, Fe, F, and Zn. The second waste fraction, coarse wastes, are either discharged to the sea or stacked on the surrounding lands. This approach to waste management has resulted in the bioaccumulation of substantial amounts of heavy metals, that is, Cd, Pb, and Cu, in the tissues of marine fauna (fish, shrimps, and crabs).

5.4.3   After-Processing Waste

Phosphogypsum is a by-product of phosphoric acid and has practically no commercial application. Its tailings pollute the territories surrounding the P acid production plants (Fig. 5.6). About 15 % of phosphogypsum finds some applications in construction industry (Phosagro 2012) but in most cases it is either stacked around the production site, fills the mined-out pits, or is dumped into the sea.

A more comprehensive environmental impact assessment of the large-scale mechanical and chemical processing of phosphate ore was made by the Environmental Protection Agency (EPA) in Florida (USA). Here, the largest national phosphate reserves, covering about 75 % of the country's needs, are situated. Company Mosaic, the main owner and operator, provides the state of Florida

Fig. 5.6 Phosphogypsum piles in Russia (Photograph courtesy of the authors)

not only with a substantial income but also employment for hundreds of Florida residents. Mosaic, however, failed to elaborate a sustainable strategy of how to tackle the harsh environmental impacts of open pit phosphate mining and the subsequent mechanical and chemical (production of phosphoric acid) processing.

This situation created intensive debates initiated by local communities and environmental groups during the recent 15 years. Their concern was heightened by the common understanding that Florida's phosphate supplies will eventually run out, or reach a point when mining is no longer economically viable. What remains then with the huge devastated territories, where at least a billion tons of phosphogypsum (the by-product of chemical processing) is loaded in 25 stacks across Florida? Every ton of produced phosphoric acid generates 5 t of waste of this controversial material, which contains radioactive contamination (radium with a half-life of "only" 1,630 years), and also elevated amounts of sulfur.

5.4.4   Application (Fertilization) Pollution

So long as water-soluble fertilizers are used for food production, substantial amounts of the deployed phosphorus (between 10 % and 30 %) are, in a medium run, removed from the productive soil layers by surface water and groundwater (Fig. 5.7).

- Phosphorus substances leached from the fields by means of runoff often can be considered a waste of different categories of toxicity.
- There exist several complicated recommendations on sustainable soil management that allow reducing the runoffs and caching more nutrients in the productive layer.

Fig. 5.7 Eutrophication of Lake Winnipeg, Manitoba, Canada (UNEP 2011)

• In general, however, so long as water-soluble fertilizers are commonly used, these runoffs cannot be controlled, and none of the measures recommended by the European Union (EU) for waste management practices can be recommended for implementation.

5.4.5 Runoffs from Farms

Other examples of the badly controlled phosphorus potential reserves are poultry and animal farms. Studies show that an average animal farm generates annually about 100 t P_2O_5 contained in the manure. Comparative studies (Helsinki Commission 2011) showed that, for the "catchment" area of the Baltic Sea, the nutrient load in the form of manure stored in (and partly spilled from) animal farms is even higher than that from the use of mineral fertilizers.

Consequent to the overuse of water-soluble fertilizers and the lack of cost-effective methods of manure recycling, these fields and farms are among the badly controlled "non-point" sources of pollution, a major environmental problem not only for the Baltic Sea.

These and multiple similar examples demonstrate that currently used technologies of production of commercially used P fertilizers are highly inefficient and ecologically damaging.

Tracing the whole chain of P-rock transformation (Cordell 2008) to the phosphorus consumed by human beings and cattle, the following picture is self-explainable (see Fig. 4.1):

- The amount of pure phosphorus contained in the P rock mined in 1995 was 17.5 M tons
- From this amount, about 2 M tons was used for nonagricultural applications (Chap. 6)
- From 15.5 M tons of remaining phosphorus, about 1.5 M tons is lost during the production and transportation operations, as already described
- From 14.0 M tons of the remaining phosphorus that is applied in fertilizers to the soil, 8.0 M tons is lost to leaching, erosion, natural disasters, etc. At the same time, this loss is compensated by the P delivered in the form of organic fertilizers (manure, compost, etc.)
- From the combined 14.0 M tons of mineral and organic fertilizers, only 12.0 M tons is transferred to the cultivated plants. Losses are caused by irreversible immobilization in the soil (see Chap. 4), feeding the weeds, etc.
- From 12.0 M tons of plant uptake, only 7 M tons is transferred to food products and for feeding livestock; 2 M tons remains in the soil as plant roots and contributes to future harvests; and 3.0 M tons is lost to fires, droughts, floods, etc.
- From 7 M tons transferred to food products and animal feed, 3 M tons ends up in the food, about the same amount goes for livestock, and about 1 M tons is lost during production and transportation
- From the same amount, livestock receives 13 M tons by grazing and an additional 3 M tons as animal food. From this amount, 15 M tons concentrates in the manure, which is later used for fertilization. As this process is of low efficiency, only 8 M tons works as an organic fertilizer, and 7 M tons is spread in the environment
- From 3 M tons of P consumed by humans, 2.7 M tons are excreted and pollute the environment

Natural recycling of the more than 17 M tons of P that now pollutes natural environments back to phosphate rock will take millions of years. Thus, this is not actually a cycle, whereby the phosphorus is regenerated in real time: it is a broken chain. From 22 M tons of P in the combined amount (mineral + organic) fertilizers, only 6 M tons is consumed by humans and livestock. The rest is a loss and calls for immediate P-recycling measures (Chap. 10).

5.5   P Fertilizers: Global Price Dynamics

The price of every P-based fertilizer depends primarily on the current cost of phosphorus (measured in P_2O_5). If the fertilizer is a single-component formulation, such as SSP or TSP, then the calculation seems to be simple:

- First, calculate the amount of P_2O_5 in 1 ton of fertilizer. For TSP 0-46-0, this amount is 460 kg.
- Then, multiply this amount to the current global price of 1 kg P_2O_5 and receive the result.

However, the global, or even the local, price of P_2O_5 primarily depends upon the price of the phosphate rock and auxiliary costs, such as outlined next.

Phosphate fertilizer costs for the end-user, that is, the farmer, consist of several major elements.

1. P rock producer costs (most often, FOB, or "free-on-board," cost)
2. Transportation costs from producer to intermediary port of destination, including insurance (for Morocco P rock and P acid, Antwerp)
3. Beneficiation and blending (such as Cd removal down to European standards)
4. Transportation costs to final production site (for example, from Antwerp to Linz)
5. All loading/unloading, packaging, storage, etc., costs
6. Recently emerged global markets speculation margins (up to 400 % in 2008)

5.5.1 P Rock Producers Mining and Supplementary Costs

As shown in Table 5.1, FOB costs demonstrate strong regional dependence. If mining costs are almost equal around the globe ($8–$12/t), processing costs range between $6 and $25/t, taxes and royalties from 0 to $10/t, transportation to port from $1 to $15, and local overhead, including return on investment, bank costs, etc., from $2 to $18. Driven by global competition, producers still are not able to equalize their expenses and make their products globally competitive. Hence, possibilities of market volatility and cartel treaties are encouraged.

5.5.2 Long-Haul Transportation Costs

Table 5.2 shows that transportation and other supplementary costs increase the final price for TSP and SSP in Switzerland by exactly 103 % compared with the FOB price in Antwerp. This enormous difference with the FOB price of SSP in Morocco (~€130/t) indicates the influence of the remoteness of the production site from the customer's site. The irreversible rise of energy and raw P rock costs in the future give no doubt that in future the price for P fertilizers in Europe will grow.

In African countries, the situation with the so-called "farmer price," that is, the one paid by the farmer to the local dealer, is alarming (Gregory and Bumb 2006). Comparing the FOB price for the same P fertilizer ($/t, 2003) for USA, Nigeria, Malawi, and Zambia, these are as follows: 135.0, 135.0, 145.0, and 145.0; that is, the FOB price is practically the same for all these countries. At the same time, the farmer prices differ substantially: 226.1, 336.15, 321.33, and 336.4! Looking at the

Table 5.1 Regional deviations in the P rock production costs (Natural History Museum)

| Country | US$ per ton of product | | | | | |
	Mine operation	Processing	Taxes+ royalties	Transportation	Overhead	Total
West Africa	10.0	12.0		1.0	3.0	26.0
Southeastern USA	8.0	12.0	2.0	3.0	3.0	28.0
Morocco	9.0	6.0	9.0	2.0	6.0	32.0
Western USA	12.0	16.0	2.0	12.0	2.0	44.0
Egypt, Jordan	11.0	12.0	4.0	6.0	13.0	46.0
Syria, Turkey	11.0	8.0	10.0	15.0	12.0	56.0
South Africa	11.0	25.0	9.0	4.0	18.0	67.0

Table 5.2 Breakdown of supplementary costs for Morocco phosphate in Switzerland

Cost fraction	€/Kg of P_2O_5	€/t TSP	€/t SSP
FOB+delivery Antwerp	0.98	448.78	175.61
Cd extraction	0.06	28.05	10.98
Transportation to Basel	0.34	157.07	61.46
Granulation	0.09	39.27	15.37
Blending, packaging	0.52	241.22	94.39
Total price	1.99	914.39	357.80

price breakdown, one sees that enormous port duties and inland transfer costs ($21.7 for Nigeria and $17.5 for Zambia, which is much higher than for the inland country, Malawi, $7.82), and bag transportation in trucks (between $60 and $70 for all African countries), make up this difference. Given the economic situation and political tensions in the continent, it is hard to imagine that African farmers would ever be able to cope with feeding the growing African population.

5.5.3   *Additional Price Uncertainty Factors*

The situation with price forecast for mineral fertilizers to Europe is especially sensitive. After Finnish P rock reserves were exhausted, the Russian reserves of high-quality apatite are being fast depleted and the Arab spring events of 2012 made other possible suppliers such as Syria, Jordan, and Egypt, temporarily unavailable; thus, Europe has no alternative but to rely on the Morocco producers.

Sedimentary rock from Morocco has two drawbacks: high concentrations of Cd and of radionuclides. Excessive Cd can be removed upon delivery to Europe at extra costs and for those customers whose national regulations request this step.

As to radionuclides, this drawback is imminent. Radioactivity of phosphate rock of different origin and mined in different locations varies significantly. Phosphate rock from different locations shows the following average figures (in Bq/kg):

Florida: for U_{238}, 1,500; for Ra_{226}, 1,600; for Th_{232}, 16
South Carolina: U_{238}, 800; Ra_{226}, 4,800; Th_{232}, 78
Morocco: U_{238}, 700; Ra_{226}, 1,800; Th_{232}, 30
Kola (North Russia): U_{238}, 90; Ra_{226}, 40; Th_{232}, 90

Removal of radionuclides from the phosphate fertilizers is practically impossible. What remains is to accept this drawback, but growing concern about the risk of radioactive contamination was spurred by the Fukushima disaster and makes this alternative not simple.

5.5.4 Uncertainties at Supply and Demand Sides

In general, the number of factors that add to the uncertainty of the global price forecast is permanently growing (Natural History Museum 2012). Several major factors defining the spot price for phosphorus (most often denominated in US dollars per ton of P_2O_5) are the following:

1. Supply side. Despite many high-level statements about discovery of new large deposits, there are several basic questions relevant to the stability of supply that cannot be fully answered now:

   • How reliable are the reported data about the new deposits? It is known that the reserve evaluation is based on interpretation of rather limited exploration data.
   • What will be the reliability of the main final product cost components, as above (mining, beneficiation, processing, stock capacities, transportation, environmental fees, etc.)? Even in several operational mines these costs are volatile. For example, as reported by Reuters on July 13, 2011, the protesters blocked for several days the railway from the Moroccan phosphate mine in Youssoufa and processing chemical plants in Safi. The same week, Moroccan security forces tried to stop a third day of riots in the central Khouribga region, which is the site of the biggest phosphate mines;
   • How will those mergers and acquisitions that are constantly going on among the major phosphate producers reflect on the final price policy?

2. Demand side

   • Phosphate demand is increasing because of population growth and rising income in developing countries, resulting in a higher nutrition value of the new diet.
   • Further increasing oil prices and concerns about the safety of oil supply enhances the cultivation of so-called "energy plants" for agro-fuel production. In some developed countries the fertilizer per crop yield improves by precision farming, but the ill-planned "bio-fuel" aspirations are endangering stabilization of demand. In total, the fertilizer demand may increase to about 187 million tonnes per annum in 2030 as a result of these influences.

As a result, the global price trend can be a long-term slow price increase. The more difficult accessibility and the decreasing quality of remaining reserves will be the main reason. If the "business-as-usual" scenario prevails and no breakthrough P-recycling technologies are implemented, the phosphate prices of about 100 US$/t (without energy plants) or 120 US$/t (with aggressive energy plant cultivation) will be most likely in the year 2030. This forecast does not exclude short-term but drastic price turbulences such as that which dominated the global markets in 2008, when the FOB price for Moroccan phosphate quadrupled within a few months. It demonstrated the volatility of the P fertilizer world markets. As a result, most European farmers were not able to buy adequate amounts of fertilizers for the next season. The price for many commodities, such as wheat and maize, increased substantially. The "Peak Phosphorus" phenomenon became a topic of serious disputes.

Summing up, the production of phosphate-based fertilizers is a vital and powerful branch of the global chemical industry operating with enormous amounts of raw materials, which are steadily being depleted and are unevenly distributed around the globe. Because of strong international competition this industry is controlled by several multinational conglomerates that have essentially divided the global market among their "areas of influence." The increase of production volumes became the main profit-making tool for producers and resulted in reduction of available types of fertilizers, such as SSP. It left the farmers with fewer alternatives in choosing optimal nutrients for their individual requirements, such as type of soil, accessibility of water resources, crop species, etc. Very few new, environmentally friendly types of fertilizers are offered by the wholesalers, especially for small family farms. No wonder that, facing the permanent need to compete on the local food markets, farmers often disregard the possible harmful effects of water-soluble fertilizers on the soil and local environment.

References

Cordell D (2008) The story of phosphorus: missing global governance of a critical resource. In: Proceedings of the SENSE Earth Systems Governance, Amsterdam, 24–31 Aug 2008

Jasinski SM (2012) U.S. Geological Survey, mineral commodity summaries. http://minerals.usgs. gov/minerals/pubs/commodity/phosphate_rock/mcs-2012-phosp.pdf. Accessed Jan 2012

Gnandi K, Tchangbedji G, Killi K, Baba G, Abbe K (2006) The impact of phosphate mine tailings on the bioaccumulation of heavy metals in marine fish and crustaceans from the coastal zone of Togo. Mine Water Environ 25(1):56–62

Gregory DI, Bumb BL (2006) Factors affecting supply of fertilizer in Sub-Saharan Africa. In: Agriculture and rural development, Discussion paper 24. The International Bank for Reconstruction and Development/World Bank. http://www.worldbank.org/rural

Helsinki Commission (2011) Fifth Baltic Sea pollution load compilation. Baltic Sea environment proceedings no. 128 http://www.helcom.fi/stc/files/Publications/Proceedings/BSEP128.pdf. Accessed 15 Feb 2012

Natural History Museum (2012) Phosphate recovery: phosphorus availability in the 21st century management of a non-renewable resource. http://www.nhm.ac.uk/research-curation/research/ projects/phosphate-recovery/p&k217/steen.htm. Accessed 10 Jan 2013

Phosagro (2012) Phosagro will extract rare metals (in Russian). http://www.rbcdaily.ru/2012/10/29/
    industry/562949985020675. Accessed 22 Oct 2012
Pierce F (2011) Phosphate: a critical resource misused and running low. http://e360.yale.edu/
    feature/phosphate_a_critical_resource_misused_and_now_running_out/2423/. Accessed 7
    July 2011
Reddy F (2004) Phosphorus from meteorites. Astronomy News http://www.astronomy.com/en/
    News-Observing/News/2004/08/Phosphorus%20from%20meteorites.aspx. Accessed 15 July
    2012
UNEP (2011) Phosphorus and food production. In: Year Book 2011: Emerging issues in our global
    environment. United Nations Environment Programme (UNEP), Nairobi, pp 34–45
Vaccari D (2009) Phosphorus famine. Scientific American, 3 June 2009

Chapter 6
Nonfertilizer Uses of Phosphorus

Although quite a substantial amount of mined phosphate (P) (about 12 %: see Chap. 4) is used for purposes other than fertilization, this issue is not reviewed in detail in this book. Although the variety and intricacy of the nonagricultural applications of P bear witness to the complex and interesting chemistry of this basic element, these applications are not considered in detail here for the following reasons:

- These applications constitute a relatively small part of the current use of phosphorus.
- The forecasts indicate that the amounts used for all these purposes are growing slower than use as fertilizer, which is expected to rise at least in proportion to human population growth, some 50 % in the next 40 years.
- Prices for the ultrapure phosphates that are used in these applications depend somewhat less upon the price of the raw phosphates; the main production costs are borne by phosphate purification and other operations.
- For many of these applications, alternative chemo-technical solutions are found.

However, for the sake of the curious reader, the most important of the nonfertilizer uses of phosphorus are mentioned (see also, for example, Valsami-Jones 2004)

6.1 Phosphorus in the Elementary Form

One group of these uses is based on phosphorus in its elementary form. As has been noted before, phosphorus does not naturally exist as such on Earth. Its high reactivity quickly converts it to phosphate or other oxidized forms and, as Brandt noted when he first discovered the element, white phosphorus, when oxidized, glows with a faint blue and green tinge. It is also highly flammable and self-ignites in contact with air. These properties of elementary phosphorus (P) in its white form are the bases of some of its uses: as an additive in napalm and a component of incendiary

M. Butusov and A. Jernelöv, *Phosphorus: An Element that could have been called Lucifer*, 53
SpringerBriefs in Environmental Science 9, DOI 10.1007/978-1-4614-6803-5_6,
© Mikhail Butusov and Arne Jernelöv, 2013

ammunition and fireworks. Phosphorus is also used to produce smoke screens, in tracer bullets, and in Molotov cocktails. Early matches also used white phosphorus, which was later changed to the less reactive red phosphorus for safety reasons.

6.2 Organic Phosphorus Compounds

Another group of uses is based on organophosphorus compounds. These compounds are often organic esters of phosphoric or thiophosphoric acid, which are powerful inhibitors of acetylcholinesterases, for example, and cause havoc in the nerve signaling system. These compounds are used in nerve gases such as Sarin, Tabun, and Soman, and in insecticides such as Parathion, Malathion, and Diazinon, as well as for treatment of lice and intestinal worms. Some of the best known and widely used herbicides are also organophosphorus compounds such as glycosate, sold under brand names such as Roundup.

These compounds are also used industrially as fire-resistant hydraulic fluids, coolants, and lubricants. In general, there are organochlorine alternatives to the organophosphorus compounds, of which the C–Cl compounds are more persistent in the environment but less toxic to humans.

Calcium salts of phosphoric acids are, besides as fertilizers, also used as baking powder ($Ca(H_2PO_4)_2 \cdot H_2O$), in toothpaste, and as a food and feed additive ($CaHPO_4 \cdot 2H_2O$).

6.3 Metallurgy

Phosphorus is also used in metallurgic processes, for example, as an important component in steel production, in the making of phosphor bronze, and in copper alloys with phosphorus and fluoride. Furthermore, it is used in glass for sodium lamps and in fine china.

6.4 Detergents

Polyphosphates are used in many detergents and as a water softener. However, these applications are permanently reduced and substituted by alternative, nonphosphoric detergents to comply with regulations on wastewater pollution.

6.5 Processing of Radioactive Materials

Phosphate is also a strong complexing agent for uranium (UO_2^{2+}), which is why apatite and other natural phosphates often contain high concentrations of hexavalent uranyl. Phosphate (as tributyl phosphate) is used industrially to extract uranium from spent nuclear fuel.

Naturally, with such wide uses, some applications are growing, while others are decreasing. Two of the traditionally large uses belong to the second category, polyphosphates in detergents, that are being phased out on environmental grounds (eutrophication), and in matches, where lighters are winning market shares. Overall, the nonfertilizer uses of phosphorus are not expected to grow much compared to its use as fertilizer.

6.6 Food Industries, Cosmetics, and Hygiene

Different types of phosphates are widely used in the food industry.

- In bakeries, phosphates are used to make dough soft. Phosphates react with sodium bicarbonate and produce carbon dioxide so that the dough becomes porous and bulkier.
- In meat processing, the reduction activity of phosphates is used to react with the muscle protein, to increase pH and reduce meat humidity. Phosphates suppress oxidants and slow down oxidation reactions in meat, poultry, and seafood.
- In the dairy industry, the calcium-binding properties of phosphates make them effective in stabilization of milk products such as cheeses, yoghurts, and milk desserts.
- Monofluorophosphates are used for production of dental care products such as toothpastes.
- Phosphate-based substances are used for water softening and other hygienic applications.

Reference

Valsami-Jones E (ed) (2004) Phosphorus in environmental technologies. IWA, London. ISBN ISBN 1 84339 001 9

Chapter 7
Eutrophication

7.1 The Processes Behind Eutrophication

Phosphorus as phosphate is as important for aquatic plants as it is for crops on land, and very often it is also the limiting nutrient. For algal plankton productivity, it can almost be said that it is the limiting factor, as we mostly do not care too much about what type of plankton carries out photosynthesis and carbon fixation. Should nitrogen as nitrate or ammonia be in short supply, while phosphate is relatively abundant, blue-green algae with nitrogen fixation ability will quickly form a large part of the algal community and secure biomass production until phosphate becomes limiting.

In nature without human interference, phosphate will slowly leak from weathered minerals, go through a number of cycles between soil and plants and soil, until it is finally in inorganic form or forms a part of organic matter, and is washed out into a watercourse where it will feed plant production. The cycling of phosphate in a body of water such as a lake or an estuary, is, however, different from and more complicated than cycling in a terrestrial ecosystem. Much of the crucial part of this was described by Mortimer in the 1940s (Mortimer 1941).

The first phase is normally the uptake: whether in a plant, plankton, or macrophyte is, in this broad picture, inconsequential. Sooner or later the plant material will end up in the bottom sediment of the body of water, for example, a lake. There the organic matter will decompose and the phosphate will be released. In a "healthy" lake with oxygenated bottom water, there will also be ions of oxidized iron, so-called iron 3 [Fe(III)], or written as chemists do, Fe^{3+}. Iron in this form will unite with the phosphate and form iron phosphate, a compound with low solubility. Often, humus substances will also be part of the complex. In this way and form phosphate will be embedded in the sediment and withdrawn from circulation in the aquatic system. Temporarily, as it turns out.

On a geological time scale, however, temporarily can be a long time. In the natural situation, such accumulation of phosphates in the bottom sediments may go on for thousands of years or longer. Sooner or later, however, will come a time, because of

M. Butusov and A. Jernelöv, *Phosphorus: An Element that could have been called Lucifer*,   57
SpringerBriefs in Environmental Science 9, DOI 10.1007/978-1-4614-6803-5_7,
© Mikhail Butusov and Arne Jernelöv, 2013

the parallel accumulation of organic matter, often coupled with changes in water flow and temperature that prolong the periods of stagnant bottom water, when the bottom water turns anaerobic. Then the Fe(III) is reduced to Fe(II) (Fe^{2+}), which does not form strong complexes with phosphate and humus, and as in an explosion phosphate is released into the water column. The Fe(II) is mostly trapped by another element that has been converted to a reduced form during anoxia—sulfur—and reprecipitated as iron sulfite (FeS). This huge increase in phosphate available to the plants will cause a growth explosion, which in turn will generate so much organic matter in the sediment that continued nonaerobic conditions are all but guaranteed. Should oxygen somehow suffice to reoxidize iron and sulfur and degrade the new large amount of organic matter, the phosphate will be retrapped by Fe(III) and left until the next anaerobic episode. In the more typical case, however, anaerobia in the bottom water will continue to be a standard condition in the lake and phosphate will continue to cause excessive growth of algae and other plants, which is the so-called eutrophication.

In the natural situation, the accumulation of iron-bound phosphate in the bottom sediments mostly parallels a gradual accumulation of both organic and inorganic matter as the lake acts as a sedimentation basin for incoming river flow, resulting in an increase in the thickness of the bottom layers of the lake. The sudden increase in internal productivity coupled to eutrophication will accelerate sediment accumulation and gradually turn the lake into a marshland, where plant roots can access the nutrient in the sediment directly. Soon enough, in geological terms, the lake will become land. Therefore, with the exception of very deep lakes in geological rifts with a relatively small catchment area, such as Lake Baikal and Lake Tanganyika, lakes are geologically young and not expected to become very old. Another sort of exception is lakes in regions such as Scandinavia or Canada, where the inland ice shield now and then scrapes away soil, lake sediment, and loose rocks and "rejuvenates" the limnic landscape (Helle 2003).

7.2 Sources of Eutrophication

For inland and coastal waters affected by eutrophication, agricultural runoff is the largest source of phosphate and other elements causing and contributing to it. The reason is simply the overuse of soluble phosphate as a fertilizer, aggravated by agricultural practices such as exposing bare soil to autumn rain and winter snow. The resulting runoff will then carry away a substantial portion of the added fertilizers.

Another important source, and in many local situations the dominant one, is the discharge of municipal sewage without adequate treatment. In a distant third place overall come industrial effluents, although they also can be locally the most important. On the top of the list here are food processing industries such as slaughterhouses and fish processing plants, but of course producers of fertilizer as well as mineral extraction plants can also be huge emitters of phosphate.

A specific thing to note, in line with the process of eutrophication as just described, is that effluents with a high oxygen demand, such as those containing reduced sulfur or easily biodegradable organic substances, can trigger eutrophication without phosphate or any other nutrient actually being added. By depleting oxygen in the bottom water, these effluents can cause Fe(III) to be reduced to Fe(II) and phosphate that had been stored over millennia to be released.

An important factor with regard to eutrophication is whether there is a stagnation period during which bottom and surface waters are not mixed. In colder parts of the world, this is generally the case in lakes in winter under the ice. If the lake is deep enough, there may also be a summer stratification caused by a temperature and density gradient. In brackish or marine waters, different water densities and thereby stratification can also be caused by differences in salinity. This stratification affects the progression of eutrophication because oxygen deficiencies are more likely to occur in stagnant bottom waters.

7.3   Eutrophication of Coastal Seas

7.3.1   The Baltic

The Baltic Sea is the world's largest body of brackish water. Nine nations border it: Denmark, Germany, Poland, Lithuania, Latvia, Estonia, Russia, Finland, and Sweden. The surface area of the sea is about 400,000 km^2, and the catchment area, with a population of 85 million people, is four times as large.

The average depth is some 50 m and the maximum depth 459 m. It is connected to the North Sea via three openings called the Danish sounds (although one has Sweden on one side). These sounds are relatively shallow, with a depth of 18 m or less. Thus, the entire Baltic Sea can be seen as a threshold fjord, and this feature sets the stage for some of the special features of the area.

In the basins of the Baltic proper and some minor deeper areas in the bays of Botnia and Finland, the water is stratified with a sharp salinity gradient at depths from 40 to 80 m. The upper, low-salinity water is mixed and oxygenated from the surface and from photosynthesis. The freshwater influx from a large number of rivers also replenishes this water layer. The saline deep water, on the other hand, is stagnant. Only under special meteorological conditions with strong, long-lasting, westerly winds is saline North Sea water pushed over the thresholds in the Danish Straits. Because of its higher density it will sink to the bottom of the Baltic basins and push out the old water there. Thus, deep-water exchange in the Baltic is an horizontal process that occurs relatively seldom, to a varying degree, and irregularly.

Thus, after a fresh inflow of oxygen-rich seawater, the oxygen will gradually be consumed by sinking organic matter from primary production in the surface waters or brought to the Baltic by the rivers. If the stagnation period continues long enough, oxygen is depleted and the bottom water will turn anaerobic. This condition occurred periodically also when human influence on the chemistry and biology of the sea was marginal, but with mankind's impact, the organic load on the bottom water has increased many fold, and the stagnation time before oxygen is depleted has decreased. Thus, anoxic conditions today are much more prevalent and longer lasting than they were some 100 years ago (Hakansson and Bryhn 2008).

Anaerobic conditions in bottom waters and bottom sediments trigger reduction of iron and release of phosphate, as just described. A self-fertilizing process is induced, mobilizing phosphate that has been accumulated in the sediments during millennia.

The total amount of phosphorus entering the Baltic Sea was estimated to be 28,400 t by the Helsinki Commission in 2006. Of this, Poland accounted for 36 %, Russia for 14 %, and Sweden for 13 %. The point-source emissions of phosphate from discharges of municipal sewage, for example, have decreased considerably since the mid-1980s and continue to do so. As an illustration, the single largest source, the city of St. Petersburg emitted 2,200 t in 2004 but only 600 t in 2011. Agricultural emissions, on the other hand, despite some efforts have largely remained on the same level even though they may not have increased. Agriculture now accounts for some 70 % of the discharge of phosphorus to the Baltic, mostly through what is termed "diffuse sources." It is to be noted also that possible future measures in agriculture will affect the load on the Baltic Sea only with a significant time delay that is dependent on phosphate retention in soils and freshwater bodies.

As only a part, and a varying part, of the bottom sediments of the Baltic are anaerobic, they act both as a sink and a source for phosphorus. The sediment exchanges in both directions are much larger than the annual input from land-based sources, but the figures are uncertain. So far, anyhow, Baltic Sea sediments appear to be a net sink for phosphate, although not carrying the same proportion of the load as they were in the past.

Eutrophication has led to some long-term shifts in phytoplankton composition. The traditional spring bloom of diatoms after the breaking up of the ice now has a significant component also of dinoflagellates. In the end of the Nordic summer, in August, cyanophytes now dominate. These shifts have consequences for nutrient cycling because dinoflagellates are much more quickly decomposed than diatoms and their nutrient content thus can be recycled in the surface water. Cyanophytes have the ability to fix nitrogen and thus contribute to the nitrogen load of the sea.

Data on nitrogen fixation in the Baltic vary significantly depending on the methods used and the time when the studies were performed. Generally, figures more recently generated are higher than the earlier ones, but this difference probably reflects methodology rather than long-term trends. Short-term and interannual variation appears high, but 300,000 t/year may be a reasonable figure for a decade average. This amount could be compared to the annual load through discharges of some 900,000 t and the loss to denitrification of perhaps 1,000,000 t. The exact data remain uncertain.

The effects of eutrophication with a drastic increase in primary phytoplankton production and a decrease in primary production and area of coverage by vascular plants and attached macroalgae, together with increased areas and frequencies of anoxic sediments and bottom waters, have also had an effect on fish populations and fisheries. Here, obviously, fishing pressure has also been a decisive factor.

The most important changes from a fishery point of view is the near collapse of the cod population in the 1980s, with only a partial recovery during the last decade,

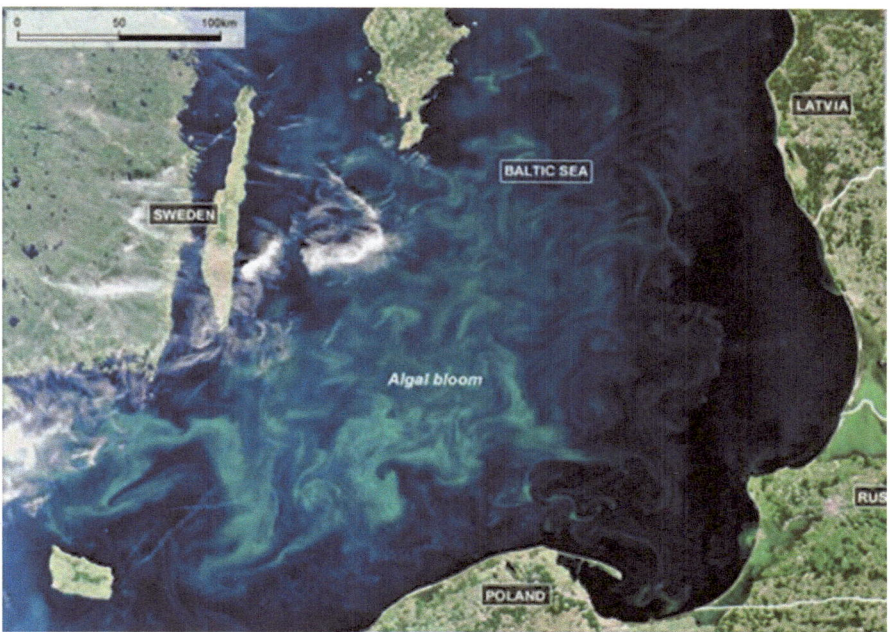

Fig. 7.1 A labeled satellite image of Baltic Sea taken on July 11, 2010, by the European Space Agency

the increase of European sprat, and the decline of herring. In addition the populations of turbot and eel are down to some 20 % of previous levels. In the case of the eel this may have little to do with the situation in the Baltic (Fig. 7.1).

The invertebrate bottom fauna has undergone remarkable changes, partly coupled to eutrophication. The population of the glacial relict amphipod *Monoporeia affinis*, once an important species in the littoral zone in the whole Baltic and a dominant species in the Bay of Botnia, decreased in numbers from the late 1970s and all but collapsed in the 2000s: it has been partly replaced by the Baltic clam. Another relict from the glacial period, the benthic isopod *Saduria entomon*, which can reach a length up to 9 cm and is an important staple for the Baltic cod, was seen as one of the first victims of eutrophication caused by low bottom water oxygen levels in the late 1960s. However, the sharp reduction in the cod population allowed it to increase in numbers in other Baltic areas and recover overall. In the middle of the 1980s, what appear to be three species of the American polychaete *Marenzellaria* came to the Baltic, probably in ballast water. Since then, it has spread explosively and is now abundant all over and totally dominant in many areas.

Populations of seals and fish-eating birds have generally recovered since the 1970s, when compounds such as DDT and polychlorinated biphenyls (PCBs), together with hunting, put several Baltic populations of the species on the brink of extinction. This loss is not directly eutrophication dependent but has an impact on overall ecosystem health.

7.3.2   Chesapeake Bay

Chesapeake Bay, the largest estuary in North America, is on the Atlantic coast with the states of Maryland and Virginia on its shores and parts of Delaware, New York, Pennsylvania, and West Virginia in upper parts of its drainage basin. The Chesapeake Bay is some 300 km long and up to 50 km wide, with a surface area of 167,000 km^2. A relatively shallow (average depth, 14 m; maximum depth, 63 m) and complex body of water, more than 150 rivers entering it, and it has a total shoreline of almost 19,000 km. Geologically, the Chesapeake Bay is a flooded river valley and thus has no fjord with a threshold.

There are some 300 fish species living in the bay and in the mouths of the tributaries, but from the fishing aspect the most important catches are those of blue crab, oysters, and clams.

Excessive discharges of phosphate and other nutrients from municipal sewage and agricultural runoff from the coastal states and those states further up in the drainage area led to increased eutrophication that gradually shifted primary production from rooted plants such as eelgrass to phytoplankton (Kemp et al. 2005). From the 1970s and onward, "marine dead zones" have been reported where anoxic conditions kill fish and other organisms. It is estimated that some 75,000 t of bottom-living clams and worms are killed yearly by anoxia, reducing the food base for the blue crabs, for example. The blue crabs themselves seem able to largely avoid the hypoxic waters by crawling onto the shores, where they occasionally amass at a so-called crab jubilee.

The oyster beds that covered nearly 1,000 km^2 50 years ago, containing well over a billion adult oysters, have been reduced to a mere 160 km^2 with some 180 million individuals today, despite recent reclamation attempts. The reason for the decline is both eutrophication and overfishing. Filtering of the bay water by oysters and bivalve mussels reduces the phytoplankton population and reduces the risk of blooms. That effect, obviously, is much weaker today.

Since the late 1990s, a toxic dinoflagellate belong to the blooming phytoplankton has caused fish mortality and rashes on swimmers' bodies.

Some measures have been taken in past decades to reduce phosphate load to the bay, but the steady inflow of some 100,000 new inhabitants to the area every year compounds the problem. Some improvements have been reported as a result of these efforts, but in January 2011 another massive fish kill occurred, with more than one million fish reported dead.

7.3.3   The Seto Inland Sea

The inland sea separates three of Japan's main islands: Honshu, Shikoku, and Kyushu. It has two outlets to the Pacific in the east and one to the Japanese Sea in the west. The length from east to west is 450 km and the north–south width ranges

Fig. 7.2 The Inland Sea with its major straits (From Wikipedia)

from 15 to 55 km. It is a relatively shallow body of water with an average depth of 37 m and a maximum depth of 105 m: it is divided into four major and some major basins connected through passages that are more shallow and narrower. In the middle section, along the Honshu shoreline, there is an archipelago with a few thousand, mostly small, islands (Fig. 7.2).

Throughout the summer, the water is strongly stratified in the main basins, with cooler and slightly more saline bottom water. The tide-dependent water exchange through the outlets does not reach the deeper basins during periods of stratification but only affects the surface water and the straits (Imai et al. 2006).

The Seto Inland Sea has traditionally been an important communication and transportation route within Japan, with extensions also to Korea and China. It has also historically been a productive fishing ground and is today a center for Japanese aquaculture producing everything from pearls to seaweed, fish, and mollusks.

The development in the Seto Inland Sea has all the ingredients of a textbook eutrophication story. Increase in levels of phosphate triggered a higher production of phytoplankton, but different types from those that previously were dominant. This change led to other phytoplankton upon which the zooplankton fed, and that change in turn favored other species of fish. Most of these other types of fish are less attractive to humans and thus of less value per kilogram.

A further increase in phosphate concentrations leads to more toxic algae and to oxygen deficiency when the excess organic matter is decomposed. At this point, fish production may decrease. A return to lower phosphate levels does not necessarily lead to a return to the original situation. Often, a new ecological order is established, with new top predators, for example.

After the Second World War, nutrients from agricultural runoff and domestic sewers made the Inland Sea highly eutrophic. This alteration initially boosted fish production but also led to a change in dominant species from red sea bream before 1960 and anchovy from 1960 to 1990. During the 1970s and 1980s, a series of measures were undertaken to reduce the discharges of phosphate to the Inland Sea. This attempt paid off insofar as the level of phosphate and total phosphorus dropped to half, but the level still remains significantly higher than pre-1960 levels. In those earlier days the algal blooms consisted mostly of diatoms. With eutrophication followed a shift to harmful raphidophytes, causing toxic red tides. During the three decades of intensive eutrophication some 300 red tide events were recorded annually, many of them resulting in extensive fish kill both among wild fish and among fish such as yellowfin tuna in aquaculture. The reduction of phosphate discharges and levels in the Inland Sea was followed by another shift in phytoplankton species composition and among marine predators. The red tide events dropped to some 100 a year, but now a different harmful and potentially toxic group of algae is dominating, the dinoflagellates. Among predators, jellyfish now take a prominent place, consuming a lion's share of the phytoplankton. Fish catches are down, and fishermen and their organizations have started to talk about "oligotrophication," believing that phosphate reduction has gone too far.

The yearly catch of wild fish was of the order of 200,000 t before 1960 and more than doubled in volume with eutrophication in the period 1960–1990, with a peak of 462,000 t in 1982. The total value, however, did not change very much because the new dominant species was of lower value. After 1990 the catches gradually dropped, reaching 256,000 t in 1999 and stabilizing at about a quarter of a million tons since then.

In addition to the wild catch, well over 300,000 t is produced annually in aquaculture. To feed those fish, a large quantity of small immature fishes is usually caught in the Seto Inland Sea, resulting in recruitment overfishing for many species. Some years, fish kills in aquaculture caused by red tides have been high, affecting especially yellowfin tuna. Those incidences have decreased following the reduced eutrophication.

7.4   Lake Eutrophication and Methods for Restoration

7.4.1   Dredging

Lake Trummen is a lake in a forest region in the south of Sweden close to the town of Växjö, currently with around 50,000 inhabitants. It is a small and shallow lake with a surface area of 1 km^2 and a maximum depth of 2.5 m. Until the early 1900s it was oligotrophic, but as the town expanded, it became the recipient of some of the municipal sewage and some industrial wastewater, which resulted in progressive eutrophication. Intensive blooms of blue-green algae occurred regularly in the summers, and fish kills caused by oxygen deficiency were a recurring winter phenomenon.

In 1958 the sewage and wastewaters were diverted, but the hoped-for improvements did not occur. The process of eutrophication and self-fertilizing of the lake was too far gone. A restoration research program under the auspices of the Institute of Limnology at the University of Lund was carried out during 1970–1971.

Through suction dredging, some 60 cm of black and brown mud, rich in phosphate and colored by iron sulfide, was removed from the lake bottom. It was pumped through pipelines to special deposition ponds, which covered an area of 0.19 km^2. The total cost of the project was just over 2.5 million SEK.

The project was successful, resulting in a significant increase in water transparency and a reduction of levels of phosphate of about 90 % and of nitrogen of about 80 %. Summer blooms of blue-green algae and winter oxygen deficiencies stopped. Improved light conditions allowed submerged plants, so-called macrophytes, to colonize the lake. Among those were blunt-leaved pondweed and stonewort. As a sign of the greatly improved conditions, the swan mussel returned in the late 1970s after having been absent since the 1930s.

7.4.2   Aeration/Oxygenation

Cardiff Bay is, insofar as the surface water is concerned, a freshwater lake around the former docks south of Cardiff center. The bay has a surface area of 2 km^2 and a maximum depth of 7.5 m. It is supplied by the Rivers Taff and Ely. It is connected to the sea through three locks. During high tide, seawater may enter and will then, because of its higher density, form a bottom layer of salty water and stratify the bay. Significant runoff of municipal and industrial sewage, especially in the form of overflow connected with heavy rains, contaminates the water and depletes oxygen.

To maintain oxygen concentration in the bay at or above 5 mg/l, compressed air is pumped, through a series of steel-reinforced rubber pipelines, from five compressors at different sites around the bay, to some 800 diffusers in the bay itself and in the mouths of the rivers. When this does not suffice, liquid oxygen from a mobile barge is added. The vaporized gas is then added to a stream of water pumped from the bay and returned after oxygenation. As much as 5 t oxygen can be supplemented this way within 24 h.

In the Cardiff Bay case, as in many others, air and oxygen are supplied to prevent oxygen depletion from intermittently ongoing discharges. In other cases, external loads may no longer be the reason for oxygen deficiency, fish mortality, high nutrient concentration, and plankton blooms, but rather the self-fertilization processes of advanced eutrophication, and aeration is used as a way to halt this process. The key then is to prevent anoxic conditions in the bottom water, which mostly occur when the water is stratified by differences in density caused by differences in temperature or salinity.

There are several ways of achieving this prevention. The most straightforward one is some version of the liquid oxygen barge in Cardiff Bay, where bottom water is pumped up, oxygenated, and returned without significantly changing the temperature: this will lock in phosphate in Fe(III) complexes and keep it in the sediment. At the same time, degradation of organic matter will increase only as a function of higher oxygen levels, not of higher temperatures, and the bottom water may act as summer habitant for coldwater species.

Another method is to bubble air through diffusers at the bottom, as in the Cardiff Bay compressor application. Some oxygen will dissolve from the bubbles themselves into the water, and much more so if the bubbles are small, as their surface area in contact with the water then is much larger than it is if the bubbles are big, and also because the smaller bubbles rise so much more slowly through the water and thus have longer contact time. However, a higher pressure is required to press the air through finer diffuser outlets, and more clogging problems with these outlets will occur. Also, the main oxygenation effect of the aeration is not through the transfer of oxygen to the water from air bubbles, but from breaking up the stratification, taking oxygen-poor bottom water to the surface, where it contacts the overlying atmosphere or light-dependent algae and higher plants can oxygenate it. As bottom water raises, warmer oxygen-rich surface water replaces it. When it comes to breaking thermal stratification, larger bubbles are more effective than smaller ones.

The question of positioning of the diffusers with regard to distance from the lake bottom is also important. If they are at the bottom, the bottom sediments will be transported up with the bubbles and water current they create, which in turn might muddy the surface water with negative consequences. If the diffusers are too far from the bottom, water stratification may persist below them, so that the aeration only changes the depth of the stratification without having much effect on the oxygen levels of the sediments.

7.4.3   Other Methods for Destratification

Besides aeration, several other methods for destratification are being practiced. Most common is mechanical stirring. Here some propeller-like devise is positioned to create a water flow strong enough to break up the stratification and cause the entire water mass to mix, leading to better oxygenated, but warmer, bottom water. The mechanically derived water current could either push bottom water up from below or down from above. The latter reduces the risk of stirring up the sediments.

Another way of creating a flow of oxygenated water from the surface to the bottom is to cool surface water and create a downstream flow. This method has several advantages in that the sinking water, if it is even cooler than the ordinary bottom water, will fan out over the deep part of the lake, oxidize the sediments without warming them, and just push the thermocline higher up without breaking it. Especially, if one of the aims is to safeguard cold-water species of fish, this is a superior method. As the needed quantity of cold sinking water also can be much better regulated to need, this is mostly a less expensive technique for the purposes than those already mentioned.

A very innovative, and in the long run the least expensive, way to prevent stratification to develop in lakes with a reasonable flowthrough of water was proposed and tried out by a Polish limnologist in the 1950s. Through a pipe, called an Olszewski tube after the inventor, hypolimnic bottom water is siphoned off to the lake's outlet, replacing the surface water that would otherwise be withdrawn from the lake (Olszewski 1961). In this way, no stratification is formed after the circulation periods.

The experience of using aeration or destratification as a way to combat eutrophication has met with considerable variation in success. The attempts to use it as a short-term measure for a year or two, hoping thereby to reduce phosphate release from sediments and the subsequent production of organic matter to the point where the oxygen of the bottom water would suffice to break it down, as it did when the lake was "younger," have largely failed. The reason is that the biological oxygen demand needed to degrade the organics and the chemical oxygen demand required to oxidize reduced forms of iron, sulfur, and nitrogen that have accumulated over longer time periods, in mostly deep layers of sediment, in most cases is simply too great. After some time, which could be some years, the anaerobic bottom conditions return and with them the self-fertilization of eutrophication.

7.4.4 Chemical Precipitation of Phosphate

In the chemical step of a modern municipal sewage treatment plant, phosphate is precipitated with Fe(III) or aluminum ions. In contrast to Fe(III), aluminum ions cannot be reduced and thus keep their phosphate-binding capacity under anaerobic conditions. This step was used in a research restoration project carried out in 1968 in Lake Långsjön to the south of Stockholm by the Swedish environmental research organization IVL.

The lake is about 2 km long but on average only a few hundred meters wide, with the maximum depth barely more than 3 m. In the early part of the twentieth century, it was the recipient of municipal sewage water, and also after the diversion of its direct outlet, overflow sewage waters after heavy rains would reach the lake. In the 1950s and 1960s it was severely eutrophicated with anoxic winter conditions regularly associated with fish kills, the foul smell of hydrogen sulfite at ice breakup, and blooms of a succession of algae throughout summer, with blue-greens forming floating mats in August. The fish population was relatively small as a consequence of winter mass mortalities.

In the early summer of 1968, 10 t aluminum sulfate granulates was sprayed over the lake. The immediate results were impressive. Visibility increased practically at once from a depth of some 40 cm to the bottom (3 m), and phosphate and total phosphorus levels in the water dropped 95 %. Also, throughout the summer, the situation was much better than previously and no real bloom of blue-green algae occurred in August. The phosphate levels steadily increased, however, but not very rapidly. During winter, oxygen concentrations under the ice dropped to low levels, but anoxia was limited to two small deeper areas.

The following spring, fish reproduction was very successful, resulting in large numbers of young fish seen in the early part of the summer. A total unexpected event also took place. For the first time ever, the freshwater jellyfish, *Craspedacusta sowerbii*, appeared in Lake Långsjön, and it did so in huge numbers. At the peak in late July and early August the population of medusas, 2.5 to 3 cm in diameter, was estimated at 100 million. The combination of the jellyfish and the small fish yearlings reduced the zooplankton population to a few percent of previous levels and the phosphate concentrations kept rising. As a result, August saw an algae bloom, but this time of green algae. The new biomass from that bloom together with the biological and chemical oxygen consumption from the sediments and an unusual amount of sewage water overflow during sudden periods of snowmelt during the winter resulted in anoxic conditions in the bottom water in roughly half the lake during the winter ice cover. Levels of phosphate and total phosphorus in the water continued to increase both from leakage from the sediments and from recurring events of sewage water overflow. Two years after the chemical precipitation the signs of eutrophication were all back, albeit in a milder form than before.

A number of other attempts were made later to combat eutrophication with chemical precipitation of phosphate, sometimes as a single remedy, sometimes in combination with other methods. The experience is mixed, but as a general pattern the method seems to have larger chances of success when eutrophication is in an early stage and the oxygen demand of the sediments is still not overwhelming. Also, deep lakes or fjords with clear and stable stratification are better candidates for this reclamation method than shallow, nonstratified water bodies.

References

Hakansson L, Bryhn AC (2008) Eutrophication of the Baltic Sea. Springer, Berlin

Helle K (2003) The Cambridge history of Scandinavia. Prehistory to 1520. Cambridge University Press, Cambridge

Imai I, Yamaguchi M, Hori Y (2006) Eutrophication and occurrences of harmful algal blooms in the Seto Inland Sea, Japan. Plankton Benthos Res 1(2):71–84

Kemp WM, Boynton WR, Adolf JE, Boesch DF, Boicourt WC, Brush G, Cornwell JC, Fischer TR, Glibert PM, Hagy JD, Harding LW, Houde ED, Kimmel DG, Miller WD, Newell RIE, Roman MR, Smith EM, Stevenson JC (2005) Eutrophication of Chesapeake Bay: historical trends and ecological interactions. Mar Ecol Prog Ser 303:1–29

Mortimer CH (1941) The exchange of dissolved substances between mud and water in lakes. J Ecol 29:280–329

Olszewski P (1961) Versuch einer ableitung des hypolimnischen wasser aus einem see. Verh Int Verein Limnol 14:855–861

Chapter 8
The Politics of P

8.1 International Politics of P

8.1.1 Guano and a Virtual Peruvian Monopoly

Farmers all over the world had long known how to put animal and human urine and feces back on their fields to increase yields. It can therefore be seen as a small step to collect also more or less fossilized droppings from birds and bats for the same purpose. Off the west coast of South America, birds feeding on anchovetas and other small fishes in the rich upwelling zones nest on a number of small islands, where guano deposits have accumulated over centuries or millennia and reached thicknesses as great as 50 m, as on the islands of Chincha. The first to utilize those and apply guano on their fields were Andean Indians. In small boats and further on, on the back of llamas, they transported the fertilizer from some of the nearby islands up to their terraced farmland in the mountains. The Incas had an advanced system for distributing the fertilizer to villages in relationship to the taxes they paid and the labor days they provided to the central system. To take guano allocated to another village was a serious crime with stiff penalties. Disturbing the bird colonies during the nesting season was also unlawful.

The conquistadors and other colonizers who ransacked the Indian societies in their hunt for precious metals and other valuables paid no attention to the ill-smelling cargo of the small vessels and pack animals. It was not until the German scientist and discoverer Alexander von Humboldt made his famous travel through the Americas around 1800 that any Westerner became interested in this stuff. He noticed the small guano boats and wrote: "You can smell them a kilometer away. The people sailing them were accustomed to the smell, but we couldn't help sneezing when they approached." Thus having become aware of it, he went on to study the substance and its fertilizer properties in the area around Callao in present-day Peru. Together with the thousands of other samples, von Humboldt brought pieces of guano back to Europe and gave parts to two French chemists to analyze.

M. Butusov and A. Jernelöv, *Phosphorus: An Element that could have been called Lucifer*, 69
SpringerBriefs in Environmental Science 9, DOI 10.1007/978-1-4614-6803-5_8,
© Mikhail Butusov and Arne Jernelöv, 2013

They found that the Chincha guano contained high concentrations of phosphorus as phosphoric acid (8–12 %) as well as nitrogen and potash.

The only phosphorus- and nitrogen-containing soil additive available at the time, besides fresh urine and feces, was bone meal. Bones from slaughterhouses routinely went to grinding factories in Britain, France, and Germany, but with the widespread depletion of arable soil in Europe, that supply did not suffice.

The desperate search for bones to grind led to the mass graves on battlefields including the fresh ones of the Napoleon wars, as well as to the cemeteries. "It's now ascertained beyond a doubt, by actual experiment on an extensive scale, that a dead soldier is a most valuable article of commerce," the *London Observer* wrote in 1822.

From this perspective, one would have thought that guano, with twice the concentration of phosphorus, would have been an immediately attractive alternative, but the recommendations of von Humboldt and his associates largely went unheard.

It was not until 1840, when Justus von Liebig published his book *Organic Chemistry and Its application to Agriculture and Physiology*, in which he strongly spoke out in favor of guano and against bone meal,[1] that the large landowners started to listen. von Liebig was highly respected in the European agricultural circles for having pioneered crops such as potatoes and maize.

A frantic guano hunt broke out. In 1841 Britain imported just under 2,000 t of Peruvian guano from the Chincha Islands. Two years later the quantity had more then doubled, and then it exploded: in 1845 it was well over 200,000 t. The crops on guano-fertilized lands doubled or tripled, and Europe took a major step away from mass starvation and toward economic prosperity, thus marking the start of the "Green Revolution."

Back in Peru, the government realized it had a prized commodity at hand and, to maximize the profits from it, the Chincha Islands were nationalized in 1842. However, as the government refused to pay more than minimal salaries, they could not find workers to mine the guano under the appalling prevailing conditions. Instead, the islands were stocked with convicts, army deserters, and African slaves, but this arrangement quickly ran into trouble. Large fights broke out between convicts and deserters and resulted in large-scale killings in addition to the toll of accidents and diseases. The owners of slaves refused to lend or hire them to the guano service on the Chinchas as they had more profitable use for them on the mainland.

After 7 years, in 1849, the Peruvian government privatized the Chincha Islands and awarded an exclusive concession to the Peruvian Domingo Elias. He had been the prefect of Lima and had declared himself to be the ruler of the nation. His claim at first was not accepted, but because he had huge landholdings and was the largest cotton producer and one of the largest slave owners, he was once in a while granted privileges to stay at bay. His Chincha concession can be seen as one of those privileges.

As part of the concession he was expected to use his own slaves to mine the guano, but he found he could not spare them from the cotton fields. He then obtained

---

[1] It was, however, not the phosphorus content of guano that von Liebig emphasized, but that of nitrogen. He might have been wrong in that, but right about guano all the same.

another gift from the government in the form of subsidies to merchants who brought in immigrants. Not surprisingly, he himself quickly became the most prominent of the labor importers through his agents in Fujian, China.

This was a time when Chinese workers were lured to the Americas on a large scale, to lay railroads in the United States, to plant and harvest cotton, coffee, and sugar in Brazil and in the Caribbean, to mine copper in Chile and Bolivia, and, among other things, to slave on the Elias' guano islands in Peru under the supervision of his African slaves. The working conditions were bound to the strict production quotas at the peril of stiff punishment and became totally inhuman. A series of reports from the Chincha Islands in the media during the 1850s created an international scandal that forced the Peruvian government to buy back the concession from Elias.

After an openly bribed competition, the Chincha concession was given to another Peruvian landlord. Elias, however, was by no means pleased with the settlement he received and sought to regain his profitable concession. He cried foul about the official corruption and tried two coups d'état that failed. When both failed, although nobody tried to prosecute Elias for his attempts, he decided to run for president but failed to win the 1857 election.

The enormous profitability of guano was obvious to many. Other influential Peruvians leased other guano islands with or without exclusive governmental concessions. They, too, quickly became superrich guano barons with an internationally noted extravagant lifestyle.

Besides granting exclusive guano mining concessions at home, the Peruvian government also yearly gave an exclusive right to a shipping company to bring guano overseas. The main benefactor of this was the British firm of Antony Gibbs & Sons of London. The Gibbs had been merchants in Lima since Spanish colonial days and knew how to strike business deals there. They signed their first guano trading contract with the Peruvian government in 1842. With the exception of a few years they maintained that monopoly for two decades.

The green revolution that guano caused in Britain in the 1840s and 1850s did not go unnoticed in Germany, France, and the USA, and a scramble for guano resources broke out.[2]

The British government tried to have Antony Gibbs & Sons give priority to British buyers, but even when that happened, the company in turn did not want to forgo the extra profits that could be had from selling it to American, German, or French farmers instead of British.

Initially, the American farmers can be said to have lost most from the rapid increase in agricultural productivity in Britain. At the same time, as they started to see their own yields decrease, their export market shrank and prices dropped. The small stream of guano that reached the United States in the early 1840s swelled to a river in the early 1950s as the prices the farmers there were willing to pay were

---

[2] An important reason for the guano race was also the use of saltpeter (potassium nitrate) from it, in the production of gunpowder, but this is outside the scope of this book.

higher than elsewhere, but the quantity that ended up on American fields was still much smaller than the amounts of guano used on British fields.

8.1.2   The Guano Island Act and Some of Its Consequences

To escape what they saw as a guano squeeze and a de facto embargo, the U.S. Congress in 1856 passed the Guano Island Act (Skaggs 1994), a remarkable piece of legislation giving American citizens the right to seize guano islands anywhere in the world and claim them as personal property and as American territory. The closest comparisons may be the royal edicts in Britain, France, and Spain that made pirates such as Francis Drake privateers by giving them a legal right to capture, for their own benefit, ships that wore the flags of enemy warring nations.

The text of the Guano Island Act may not seem that alarming. The first section reads: "Whenever any citizen of the United States discovers a deposit of guano on any island, rock, or key, not within the lawful jurisdiction of any other Government, and not occupied by the citizens of any other Government, and takes peaceful possession thereof, and occupies the same, such island, rock, or key may, at the discretion of the President, be considered as appertaining to the United States."

The devil is, so to say, in the interpretation. First, "discovers" does not mean that it should be an earlier unknown place. It could well be a guano island that had been known for centuries and mined for decades. "Discovers," in this context, is best understood as "sets eyes on." Second, "lawful jurisdiction" means that the USA has officially accepted that the island belongs to a specific country, which at the time was the case with very few outlying islands. Third, "not occupied" meant "not permanently inhabited," which hardly any guano island was as they typically lacked both drinking water and vegetation.

Fourth, "take peaceful possession thereof" did not exclude evacuating people from them with force.

Once a claim had been made and the president had acknowledged the annexation, it would be the task of the U.S. military to protect the interest of the "discoverer." Is this surprising at all nowadays?

Let us take a few examples to see how awkward this was sometimes played out.

Navassa Island is situated in the Caribbean off Haiti. In 1504, when Christopher Columbus was stranded on Jamaica, he sent some crew members to Hispaniola for help. Traveling by canoe they ran into the island, but soon went off, as they found no freshwater there. Haiti claimed the island in 1801 and wrote it into its constitution after final independence from France in 1804. All the same, Peter Duncan, an American sea captain, claimed it for himself and for the USA under the Guano Island Act on September 19, 1857. Haiti protested, but the U.S. President James Buchanan in July 1858 issued an Executive Order upholding Duncan's claim and calling for military action to support it. In a ruling in November 1890, the U.S. Supreme Court found that Navassa Island was to be considered "as appertaining to the United States."

Today, Navassa Island, long since depleted of its guano, is administered by the USA as an "unorganized unincorporated territory" against the sustained claim of Haiti.

Bird Island, also in the Caribbean, is situated off Venezuela and west of the Leeward Islands. The island was first discovered by a Spanish adventurer, Avaro Sanzze, in 1587, who claimed it for Spain and named it Isla de Aves (Bird Island). In 1678, the island became reputed as dangerous when a whole fleet of 17 French vessels under the command of Marshal Comte d'Estrées wrecked on the reefs. The island was later to change possession among the British, the Spanish, the Portuguese, and later the Dutch, who all in turn concurred and claimed it. At independence from Spain, first as part of the Republic of Gran Colombia in 1821 and then as a separate sovereign country, Venezuela claimed Isla de Aves.

In 1854 an American captain noted the presence of guano on the island and started to exploit it some time thereafter. He also later claimed it under the Guano Island Act, but the U.S. president never acted on the claim. The Dutch, however, did and sent a man of war there to clarify their position and demanded that the Americans obtained a Dutch permit before continuing the operations. Such a concession was granted, provisionally, by the Administrator of St. Eustatius in 1859, to be confirmed by the Governor of Curaçao. The holder of the permit was Edward Green, Kean & Co., in Baltimore, a company in which the Sanford family had a dominating interest.

All along, Venezuela had protested and repeated its claims to the island. While the permit question was still being negotiated, Venezuela and the Netherlands agreed to an arbitration of their respective claims, and Isabella II of Spain was chosen to decide the fate of the island. She ruled in favor of Venezuela, but with unlimited Dutch fishing rights in the area. The Dutch accepted.

However, fearing that the U.S. president would accept to have Bird Island annexed under the Guano Island Act and send the U.S. Navy to protect the interests of American guano miners, Venezuela did not dare to fully enforce the nullification of the Dutch provisional concession. So, American guano exploitation at Isla de Aves occurred sporadically until 1878, when another U.S. company, the Guano & Copra Company of America, started mining operations that continued until 1912, when the bird manure was exhausted for all practical purposes.

In the meantime, Henry Sanford had started a campaign to obtain compensation from the Venezuelan Government for the nullification of the concession to the Isla de Aves guano that had been held by his uncle and other businessmen from Baltimore. His legal and diplomatic maneuvers took decades, but after political favors granted by the Americans, and what his biographer Fry describes as "covert lobbying" and "well-placed bribes," he made a fortune when Venezuela generously settled his claims to Birds Islands guano

So, in the end, the United States took the guano from Isla de Aves, Birds Island, with the threat of, but without enforcing, the Guano Island Act, and one American received a sizable sum of money from Venezuela for not being the one who took the guano.

Isla de Aves or Birds Island is today Venezuelan territory and is not claimed by the USA.

The main part of the claims and annexations under the Guano Islands Act, however, happened in the Pacific. One such area was the *Johnston Atoll*. It was first reported seen in 1796 when a brig named Sally under the command of Captain Joseph Piermont stranded on a shoal not far from the island. He later gave the accurate position of the atoll, but neither named nor claimed it: that took place 13 years later when Captain Charles Johnston on the Royal Navy's HMS *Cornwallis* arrived there.

Despite this British claim, in 1858 both the Kingdom of Hawaii and the United States of America claimed the island, the latter under the Guano Islands Act. Following what is euphemistically called the Reciprocity Treaty of 1875, the island and atoll came under formal U.S. jurisdiction. By then guano mining by the Americans was well under way and by 1890 the deposits had been depleted.

Later, in the 1950s and 1960s, the USA turned the atoll into a testing site, both atmospheric and underground, for nuclear weapons and a storage and destruction facility for chemical weapons.

Sometimes the interests and ambitions of the Kingdom of Hawaii and the USA collided, as it happened concerning the *Palmyra Atoll*. It had been regularly visited by Hawaiians for generations when it was first sighted by Westerners. The American captain Edmund Fanning, on board his ship *Fanny*, narrowly escaped stranding on the atoll in 1798 on his way to Asia. The first Westerner to actually set foot on the island had less luck. In November 1802, the American ship *Palmyra* with Captain Sawle wrecked on the island.

In 1858 Dr. Gerrit Judd on the brig *Josephine* claimed Palmyra for himself and the USA in accordance with the Guano Island Act, although the amount of guano on the island was too small to be of any commercial interest. The Hawaiian King Hamehameha IV in response sent his navy and two emissaries, the Hawaiian citizens Bent and Wilkinson, to take possession of the atoll, which they did, and in April 1862 it was formally annexed to the Kingdom of Hawaii, with the U.S. claim lying dormant. The private ownership then changed hands multiple times after it was sold by Bent and Wilkinson. On the governmental level, the situation became more complicated when the British Commander Nichols on HMS *Cormorant* formally claimed Palmyra for Great Britain, ignoring the prior claims of Hawaii and USA. (The latter could be argued on the ground that the conditions for application of the Guano Island Act had not been in place.) The USA then annexed Palmyra in conjunction with the overall annexation of Hawaii in 1898 and made it part of the Territory of Hawaii in 1900. To counter the British claims and legal arguments once and for all, a second, special, act of Palmyra annexation was issued by the USA in 1911. When Hawaii in 1959 became a U.S. state, Palmyra was explicitly separated from it and became a U.S. federal incorporated territory.

In total, more than a hundred islands, atolls, and rocks were claimed under the Guano Islands Act, laying the ground for a significant geographic expansion of U.S. administrated territories. After depleting the guano resources, most of the islands claimed under the act have been abandoned or handed over to other states. Unfortunately, the concern for protection of the bird colonies, well kept by the Incas, was not a part of the interests of the Western industrialists, neither the

Americans nor their European counterparts. Thus, guano extraction has gone hand in hand with bird colony devastation.

The territories acquired through the Guano Islands Act that remain in U.S. possession are Baker Island, French Frigate Shoals (part of Hawaii), Howland Island, Jarvis Island, Johnston Atoll, Kingman Reef/Danger Rock, Midway Atoll, Palmyra Atoll, Swains Island (part of American Samoa, no evidence of guano extraction), Navassa Island (contested by Haiti), Bajo Nuevo Bank (contested by Colombia), and Serranilla Bank (contested by Colombia).

8.2   Rock Phosphate and the New Game

In 1840, Sir Johan Bennet Laws, the founder of Rothhamsted Experimental Station, had discovered that he could convert the phosphorus in animal bones into a soluble phosphate that can be readily used by plants. He treated bones with sulfuric acid and called the product "superphosphate." He received a patent and started marketing his product in 1842, and this can be seen as the beginning of the end for guano as the prime phosphate-containing fertilizer. Bone meal, in short supply as it was, was quickly replaced by phosphate rock as the raw material for similar processes to produce superphosphate. Although rock phosphate was more available than guano, it was still not widespread. Commercial mining of phosphate rock began in the mid-1880's. In 1847 the first 500 t had been extracted (in Suffolk, UK). By 1928 the yearly world production had reached 10 million tons and in 1974 it passed 100 million; in 2010 it was 170 million tons with China as the largest producer (Table 8.1).

Table 8.1   Major producers of phosphate rock in the world: 2010

Country	Production in 1,000 metric tonnes
China	65,000
United States	26,100
Morocco and Western Sahara	26,000
Russia	10,000
Tunisia	7,600
Jordan	6,000
Brazil	5,500
Egypt	5,000
Israel	3,000
Australia	2,800
Syria	2,800
South Africa	2,300
Algeria	2,000
Togo	800
Canada	700
Senegal	650
Other countries	9,500

The total mineable quantities, the reserves, are today estimated at 65 billion tons,[3] but the term "mineable" is economically defined, and if a significantly higher price was accepted, reserves would also be significantly higher.

The problem at this point is time and the mining site. For the decades to come the main phosphate reserves are concentrated to a few areas, especially Morocco and West Sahara, with 50 billion tons, 77 % of the total global figure.

Historically, large amounts of phosphates were obtained from deposits on small islands such as Christmas Island and Nauru, but these sources are now largely depleted. Let us see how the international phosphorus (P) politics played out there before we return to the question in relationship to Morocco and West Sahara.

Christmas Island was first sighted in 1643 by Captain William Mynors, on the British East Indian Company vessel *Royal Mary*, who sailed by it on Christmas Day. The first who actually set foot on the island were two crew members of the British ship *Cygnet* under the navigator William Dampier. At their visit and sail round the Christmas Island in March 1688, they found it uninhabited. Almost 200 years later, in 1887, the British naturalist J.J. Lister on board HMS *Egeria* spent 10 days on the island collecting biological and mineralogical samples. Among the rocks were many of "nearly pure phosphate of lime," according to subsequent analyses. On the basis of those findings, the British Crown annexed Christmas Island in June 1888. The administration was entrusted to the Straits Settlements, which meant to be run jointly by the British Phosphate Commissioners and the UK Colonial Office. In 1890, the British Phosphate Company started mining the phosphate using imported workers from Singapore, Malaysia, and China. Later the administration was transferred to the Crown Colony of Singapore.

During the Second World War, Christmas Island was concurred by Japan for its phosphate wealth in March 1942, but after the end of the war the island was reclaimed by Britain and returned to the administration of Singapore.

In 1957 the UK transferred sovereignty of Christmas Island to Australia, and the Australian government paid £2.9 million in compensation to the Singaporean government, a figure based mainly on an estimated value of the phosphate forgone by Singapore.

8.2.1   Nauru

The first inhabitants of Nauru Island were from Micronesia and Polynesia and came there at least 3,000 years ago. They were divided into 12 clans, nowadays represented by the 12-pointed star in the Nauruan flag. The Nauruan were early aquaculturists catching milkfish (*Chanos chanos*) fry in the sea and releasing them after acclimatization in the eutrophic freshwater lagoon of Buada, where they also were given additional feed. The first Westerner to visit the island was John Fearn, captain of a British whaler. He came there in 1798 and so much liked the place that he

---

[3]These figures are from the United States Geological Survey (USGS) 2010. The year before the same agency estimated the total global reserves to be 16 billion tons.

named it "Pleasant Island." From 1830 Nauru became a place where whalers regularly picked up supplies of food and freshwater, often in exchange for alcohol and firearms. These depredations were probably important factors behind the Nauruan tribal war that started in 1878 and went on for a decade, significantly reducing the population of Nauru.

In 1888 Germany annexed Nauru, ended the tribal war, invited Christian missionaries, renamed it Nawodo, and incorporated it in the German Marshall Islands. The tribal system had partly disintegrated during the war and was replaced by kings who reigned under German supervision.

In 1899, Albert Ellis, working for an Australian company that traded in Pacific phosphate, copra, and pearl shell, found that the Nauruan rock that served as a doorstop in his Sydney office was almost pure phosphate. He traveled to the island and in 1900 confirmed that Nauru had significant rock phosphate deposits. Under an agreement with Germany, the Pacific Phosphate Company started phosphate exploitation in 1906.

When the First World War broke out in 1914, Australian troops captured Nauru. Together with New Zealand and UK, Australia formed the British Phosphate Commission (BPC), which took over the phosphate mining rights. In 1923, Australia obtained a trustee mandate over Nauru from the League of Nations with New Zealand and UK as co-trustees, and BPC continued its phosphate mining. During the Second World War, in December 1940, German cruisers shelled phosphate mining and shipping areas and sank some supply ships. Japanese troops then seized Nauru in August 1942 and built an airstrip, which was bombed by the Americans in March 1943, preventing delivery of food supplies to the island. The Japanese then evacuated—or deported—1,200 Nauruan, almost the whole population, to Chuuk Islands. In September 1945, the Japanese troops on Nauru surrendered to the Australians, and in 1946, the surviving 737 Nauruan were repatriated from Chuuk. In 1947 the United Nations gave Australia, New Zealand, and Britain a trusteeship over Nauru similar to that of the League of Nations of 1923, and BRC resumed its phosphate mining operations. In 1968, Nauru became independent after 2 years of self-governance, during which time the people of Nauru purchased the local assets of BPC for 21 million Australian dollars and passed it on to the national Nauru Phosphate Corporation. The income from the phosphate mining briefly gave Nauru the highest per capita income of all sovereign states in the late 1960s and early 1970s. The production then dwindled until the deposits finally ran out in the 1980s, leaving a destroyed landscape as the historic landmark of the mining period (Fig. 8.1).

In 1989, Nauru took Australia to the International Court of Justice over its actions during its trustee administrations of the island. The legal complaint had three main components:

- The failure of Australia to remedy the environmental damage caused by phosphate mining.
- The failure of Australia to repatriate any significant part of the income from the mining to Nauru
- The forced purchase of national assets before independence and at a price determined by past profits disregarding that the phosphate deposits were almost depleted.

Fig. 8.1 A limestone karst in the Nauru after overmining of phosphate (From Wikipedia)

The case lead to an out-of-court settlement, under which Australia would reha-
bilitate certain exhausted areas of Nauru.

Morocco and Western Sahara together are the third largest producer of phosphate
rock and the largest exporter, because China and the USA, although mining more
phosphate, largely use it for domestic consumption. As stated earlier, Morocco and
Western Sahara also have by far the largest reserves. All the production is carried
out by the state-owned Office Chérifien des Phosphates (OCP) in four main mining
areas, of which one, Bou Craa, is situated in the occupied Western Sahara. It pro-
duces some 3 million tons of the OCP total of 26 million, but the stock is of the best
quality and highest phosphorus content, so in value the Bou Craa production
amounts to 11 % of OCP sales.

The Bou Craa phosphate deposits were first discovered by Spanish geologists in
1947, when Spanish Sahara was a colony. The deposits are near the surface and very
pure. They are situated some 100 km southeast of the city of El Aaiun. The mining
started in the 1960s, and the world's longest conveyer belt was constructed to take
the crushed rock to the harbor of El Aaiun. In 1973, the phosphate export from Bou
Craa and Spanish Sahara was the sixth largest in the world.

In 1975 Spain withdrew and Morocco quickly invaded most of the new Western
Sahara, including the Bou Craa area, and claimed the country for itself. It also sent
some 350,000 of their nationals as settlers with tents on their backs over the border.
The native Sahrawi fought back for 16 years under the leadership of the Algerian-
backed Polisario rebels, signing a ceasefire in 1991. After taking control of the mining area and transferring the operations to OCP, the Sahrawi workers were fired and
replaced by Moroccan workers.

In October 1975 the International Court of Justice rejected territorial claims by Morocco and Mauritania. The court recognized the Saharawi right to self-determination and Spain agreed to organize a referendum. The Moroccan occupation and the massive immigration of settlers from Morocco have led to continuous disputes between Polisario, supported by Algeria, and Morocco as to the terms of the referendum and the question: who would be eligible to vote. Despite several UN and American attempts to break the impasse, no real progress has been made. According to the UN, Western Sahara today is a "non-self-governing territory," the status of which is "disputed." Calls on Morocco to end the illegal occupation of Western Sahara and stop the illegal exploitation of the country's natural resources, notably phosphate and fisheries, have gone unheard. No doubt, the phosphate riches are obstacles to independence for Western Sahara (Zunes and Mundy 2010).

8.3 Domestic Politics of P

On the domestic level, the politics of P has concentrated on two areas, the environmental effects of phosphate rock mining and processing and national ownership of the deposits or free market access. Not only does one country accounts for 75% of the global reserves and five countries for 75% of global production, but within those just a few companies dominate, many of which are state-owned. As an example, Office Cherifien de Phosphate has a monopoly in Morrocco (and West Sahara). The Mosaic Company in the USA owns the world largest phosphate mine in Florida and is the single largest producer of finished phosphate products with more than half of the U.S. production and 13 % of that of the world. FosAgro dominates the Russian phosphate mining with large mines, that is, in Khibiny on the Kola peninsula, whereas the Yuntianhua Group of China is a Middle Kingdom big player with extensive phosphate extraction, for example, in Hunan Province. In total, ten corporations account for more than half the global extraction of phosphates.

Nature itself, of course, is the reason for the concentration of phosphate resources in a few countries, whereas the risk of losing the substantial profit from the extraction and possible access to the product itself, as could happen on an open and free market, is the reason for countries to have national state-owned or state-controlled enterprises do the mining and processing. Then, given that the phosphate-holding geological structures are few and tend to be big, the companies that exploit them tend to be large, too.

When it comes to the environmental policies around phosphate rock excavation and processing, the fact that the industries are big, state owned, and a source of significant governmental income makes them the Goliaths against the Davids of national regulating agencies in environmental as well as in other fields. The same is the case with huge national oil companies, for example: the regulators become the underdog and environment and health suffer.

Phosphate mining is often carried out as open-cast strip mining, leading to a bare world of mine tailings reminiscent of a dead moon landscape with contaminated aquifers and toxic or phosphate-overloaded runoff water. Gypsum is a by-product when phosphoric acid is produced from rock phosphate, and large volumes of it in landfills of mountainous proportions are a common sight at processing facilities all over the world. Fluorides become air contaminants that can damage both plants and livestock as well as human health. In some places nowadays it is scrubbed out before the gas is released through the smokestack; in others it is not yet. In addition to these problems, phosphate deposits mostly contain heavy metals such as cadmium and radioactive elements such as uranium-238 and radium-226. Dependent on the operation, these elements may end up in the air, in runoff, in wastes, or in the products themselves.

As can be seen from this compressed list, the environmental problems associated with rock phosphate mining are huge and, as said, the operating companies have mostly not been obliged to address them. There are recent cases, however, as in Florida when public protests and court decisions have forced the Mosaic Company to cut down on plans of expanding their South Fort Meade mine. As a *Herald Tribune* article has it (Spinner 2012):

> The battle over phosphate mining has shifted
> By Kate Spinner
> Published: Wednesday, February 22, 2012 at 7:30 p.m.
> Last Modified: Wednesday, February 22, 2012 at 7:30 p.m.
>
> When Southwest Florida local governments gave up the fight four years ago against a phosphate mine that could have threatened the regional water supply, three small environmental groups pressed on.
>   The groups' persistence paid off Tuesday in a compromise with Mosaic, the world's largest phosphate mining company, that protects more land from mining, adds buffers around streams and wetlands and preserves more than 4,000 acres near a major water utility that supplies drinking water to tens of thousands of people in three counties.
>   The settlement marks a shift in a legal dispute that has been waged for more than a decade over the extent of mining, and practices for extracting phosphate to make fertilizer, in the Peace River basin.

One can also add that in some U.S. cases the phosphate rock mining companies have been forced to reclaim mined areas in a similar fashion as Australia had to do on Nauru, as described earlier. So far, however, these are the exceptions and might remain so for a long time to come, unless the title of Kate Spinner's article has it right.

Signs of a shifting battle are so far not seen in China, Morocco, or Russia. There, the phosphate giants continue to rule—for now.

References

Skaggs JM (1994) The Great Guano rush. Entrepreneurs and American overseas expansion. St. Martin's Griffin, New York
Spinner K (2012) The battle over phosphate mining has shifted. Herald Tribune, New York
Zunes S, Mundy J (2010) Western Sahara. War, nationalism and conflict irresolution. Syracuse University Press, New York

Chapter 9
Peak Phosphorus

The expression "Peak Phosphorus" stems from the earlier works of M.K. Hubbert, a geophysicist from Shell Oil, who already in the 1960s correctly predicted that oil production in the USA would reach its peak before 2000. Further on, his method was applied to forecasting of the exhaustion time of the global oil reserves as well as gas, coal, uranium, and, which is the most interesting for us, phosphorus.

9.1 Principles of Hubbert's Linearization

This work by Hubbert was called "depletion analysis" and, in application to the oil reserves, was based on the following assumptions:

1. Oil reserves are naturally limited but, at the same time, increasingly exploited to fulfill the needs of the growing population.
2. Therefore, the time will come when these reserves are exhausted unless the proper alternative resources are found.
3. Even if such alternative resources appear, but cannot substitute for oil fully in all areas (for example, gas and nuclear power could not substitute for oil in many significant applications), it is still possible to predict the reserve exhaustion.
4. In both cases, as a result of potential exhaustion, the curve describing the time dependence of annual oil production should have a maximum (so-called Peak Oil), after which the production rate will decrease.
5. The significance of Hubbert's finding was that it is possible, even during the rising tendency of the rate of oil production, on the basis of mathematical analysis of this rising curve to predict when such a peak oil happens and when the reserves are expected to be exhausted.

It is pretty clear why the time dependence of the mining rate of a limited reserve should have a bell shape. Let us first take a simplified example of a limited reserve: a glass of water.

M. Butusov and A. Jernelöv, *Phosphorus: An Element that could have been called Lucifer*, 81
SpringerBriefs in Environmental Science 9, DOI 10.1007/978-1-4614-6803-5_9,
© Mikhail Butusov and Arne Jernelöv, 2013

Fig. 9.1 Time dependence of
the exhaustion rate of two
limited reserves

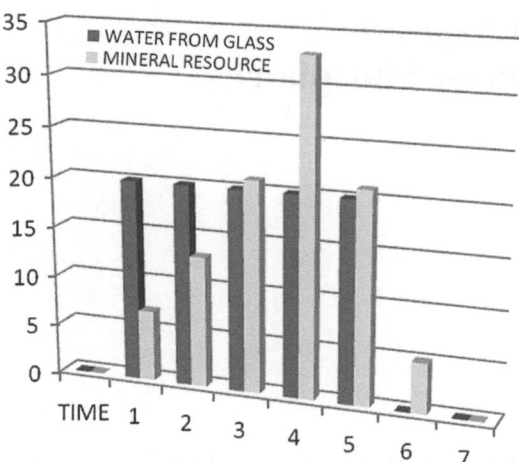

Disregarding the last drops, one may say that in the case of constant water flow poured from the glass, the exhaustion rate is described by a flat curve (dark gray bars on Fig. 9.1), which abruptly starts from zero and comes again to zero as soon as no water remains in the glass.

The situation with mineral reserves is different. After the first discovered sources of such reserves are explored, their exploitation rate grows, and this growth rate depends upon the increase of consumption (demand side) and increase of production capacities (supply side), such as follows:

- Growth of global population
- Growth of the part of the population willing and able to acquire this reserve (in the case of oil, the segment of population using cars and planes)
- Development of new oil reserves
- Development of new oil exploration technologies, etc.

Then, when the limits of the reserve conflict with the growing demand, the supply chain can no longer provide as much as is requested, and production starts decreasing (light gray bars on Fig. 9.1), unless new oil deposits can be put into production that are as rich and easy to exploit as the previous sources. If not, the production rate either drops or production becomes increasingly expensive.

Indeed, oil production started only in the twentieth century (the Diesel engine was patented in 1892, still as a machine using vegetable oil!). From then, its production rate grew faster than the population growth for reasons of expansion of different applications.

On the assumption that this bell-shaped curve is close to the standard Gaussian distribution curve, Hubbert elaborated a method of predicting the position and value of the peak, the total amount of the reserve, and its exhaustion time by analyzing only a currently available, still rising part of this curve, that is, at a point much

Fig. 9.2 Depletion curve calculated by using the simplified Hubbert's method

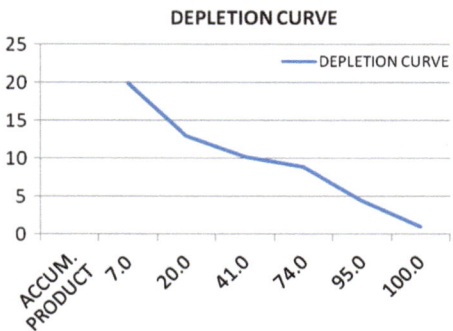

earlier than when the reserve is exhausted. The simplest way to explain how it works is a method is based on calculating and plotting, on the vertical axis, the function P/Q, where P is annual production and Q is total accumulated production up to this year. On the horizontal axis, the value Q is plotted. By this operation (called linearization), we determine a declining production depletion curve, almost a line. The reserve exhaustion time will come when this linearization curve meets the Q axis (Fig. 9.2). The point at which the curve closes the Q-line indicates the total amount of reserve that can be exploited. This amount is called the ultimate recoverable reserves (URR).

Returning to Fig. 9.1 and summing up the values of all bars, we find the amount 100: the depletion curve shows 102, which is quite accurate.

To check how this works for early predictions, one can make calculations on the basis of only data available before the peak is reached (in our case, calculating only the data from bars 1–3), that is, when accumulated production is still 41. Extending the depletion curve from there, we still come to almost the same point, 102.

This spectacular forecast method cannot be completely accurate because it disregards that the real curve may differ from the Gaussian shape and have certain temporary kinks, for many reasons such as these:

- The method is not applicable to individual countries unless they produce only for their economical needs and do not import, or if they are economically sealed off.
- Newly explored reserves can be put into operation almost simultaneously.
- Global political events (such as the Iraq war, Arab Spring, or Iranian conflict) could impede oil production en masse
- Some international conventions (such as OPEC) might decide on abrupt growth or decay in the oil supply: indeed, Hubbert came up with his theory before the OPEC interventions
- New and extremely efficient alternative sources of energy that might influence global oil demand may be discovered

Further development of the Hubbert linearization method (Deffeyes 2005) made it possible to predict the oil production rate and compare it with real data.

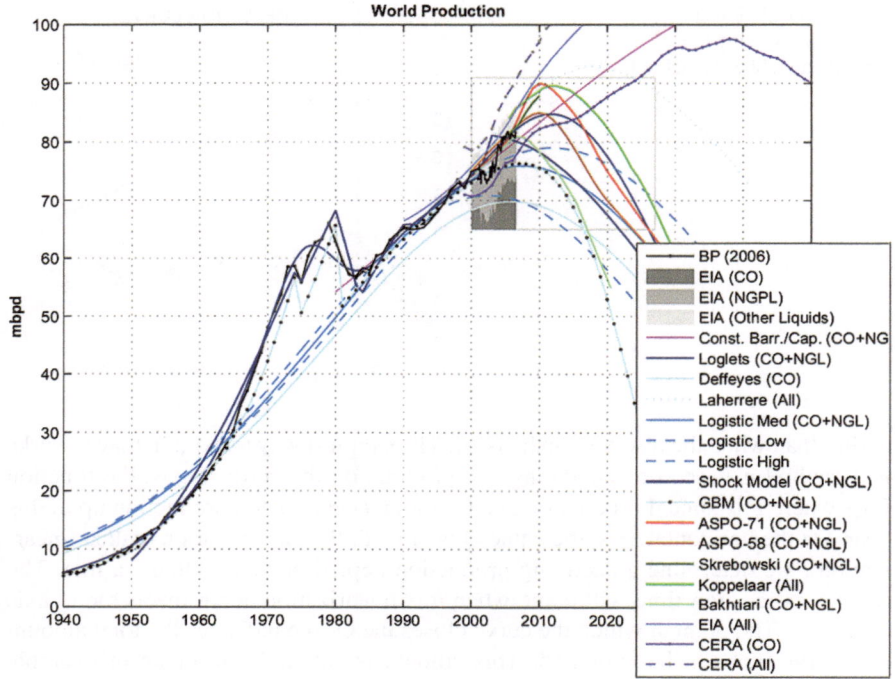

Fig 9.3 World oil production forecast (Khebab 2006)

K. Deffeyes nominated Thanksgiving Day, November 24, 2005, as World Oil Peak Day (Fig. 9.3).

What happened after, everyone knows who drives a car or travels by planes. The oil price developed a general trend of irreversible growth. Again, the definition of the Peak of Reserves did not mean that at this moment the production ceases. It meant that after the peak the production would become more expensive because deeper and smaller reserves need to be exploited to fulfill the global demand. In the same way, Deffeyes and other analysts have predicted a peak production for reserves such as natural gas, coal, and uranium.

It would be fair to say that the global community responded to the warnings on the future of oil production, although with substantial delay. However, the following measures should be mentioned.

- Development of more economical machines for transportation
- Switching part of the car fleet to gas, as a main fuel
- Overall energy protection measures, such better thermal insulation
- Development of alternative sources of energy (bio-fuel, solar, wind power, etc.)

9.2   Applying Hubbert Linearization to Phosphates

The first spectacular case of what unlimited phosphate exploration can cause was demonstrated by the tragic example of Nauru Island, where the large part of the natural landscape was destroyed within 90 years and became virtually a "moonland" (see Chap. 5).

But, for the time being phosphate markets did not perceive the idea that the same fate awaits the global phosphate reserves as well. A strong message on total exhaustion of the global P reserves was sent again in 2007 (Dery and Anderson 2007). Because different P reserves contain phosphates of different concentrations, the relevant calculations were made based only on the data for commercial phosphates (26–34 % P_2O_5). Before expanding the analysis to the global reserves, the method was tested on the known data of the Nauru reserves exploitation, but using only the data for the ascending part of production curve, that is, until accumulated reserves reached 20,000 Kt. The estimated IRR (Ultimate Recoverable Reserves) was 97,000 Kt. The calculations made on the data of the total exploration period (1959–2005) showed IRR 73,000 Kt (Fig. 9.4). In reality, the Nauru IRR was 78,000 Kt. The accuracy of the method was quite high:- the mistake was within ±20 %.

Looking at the global P production curve, it seemed evident that around the year 2005 the world had reached the production peak (Fig. 9.5)

Making Hubbert's linearization of this curve, it becomes possible to estimate the global phosphate IRR (Fig. 9.6), which seems to be 8,000 Mt. In the later

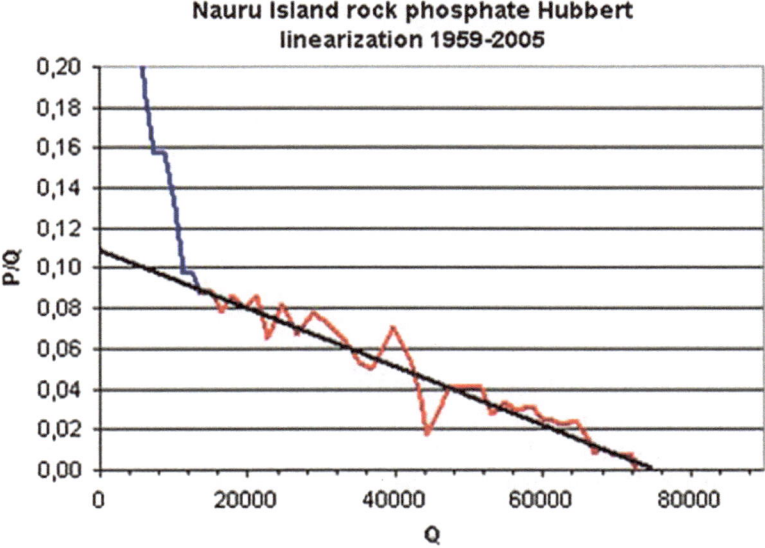

Fig. 9.4 Linearization curve based on the complete exploration history of Nauru (Dery and Anderson 2007)

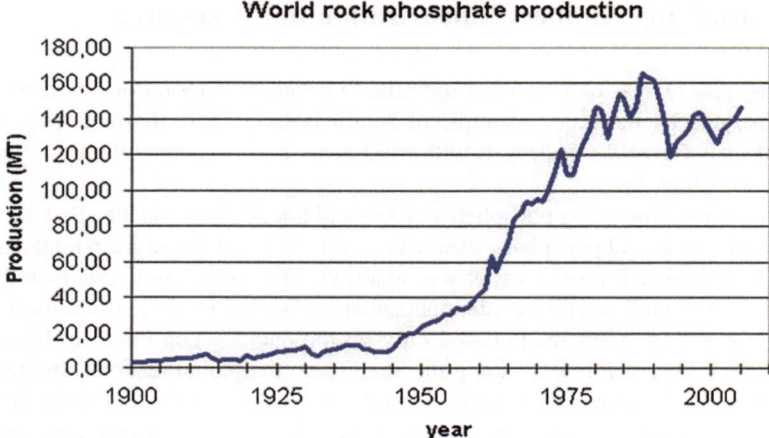

Fig. 9.5 World phosphate production curve (Dery and Anderson 2007)

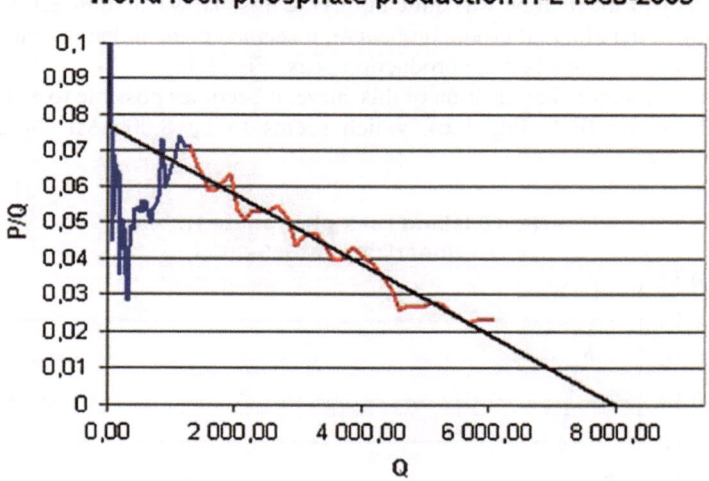

Fig. 9.6 Hubbert linearization of global phosphate production (Dery and Anderson 2007)

publication (von Horn and Sartorius 2009) this amount is about 12,000 Mt, but such discrepancies are not unusual in global forecasting.

The significance of Hubbert's approach is the ability to predict, before the first signs of trouble, both the time when the production peak can be expected and the time when the so-called "ultimate recoverable reserves" (URR) will come to an end. By connecting these two features in one forecast, analysts came to the conclusion that the trouble begins not when we "run out" of a given natural resource but when its production reaches a peak. The earlier can we predict when this peak occurs, the more time do we have for preventive measures, be it saving relevant consumption, searching for alternatives, or something else.

It seems like a coincidence, but in 2008 the phosphate rock export price FOB Morocco increased to $110/t from $60 in 2007. This kink created a panic in the phosphate markets that lasted several months until the price dropped to $80/t. The analyzers' forecast shows long-term tendency for the increase, which may indicate that we have already reached the peak for phosphorus.

What has been done to prevent the exhaustion of P reserves in the twenty-first century? So far, not much. In contrast, several factors indicate that the global demand will not decrease but is growing, for many reasons:

- The global population grows continuously, by at least 110 million people a year, and in 2012 reached 7 billion.
- Nutrition in several heavily populated countries demands more fertilizers, because the change from the soya diet to that of chicken meat requires about four times more grain for feeding the chicken than for soya, with equivalent protein value.
- Political decisions to develop energy independence from the external oil supply resulted in allocation of more lands to so-called "energy plants." For example, the European target is that in 2020 the amount of agrofuels used in transportation should be increased to 10 %. If such a trend continues worldwide, in 2030 about 13 % of the total fertilizer demand will be used for the production of energy plants.

Intensive implementation of these "energy plants" is one of many indicators that the issue of Peak Oil, although not timely recognized, finally was taken under serious consideration. Intensive research was financed by different states and companies on the development of more economic car models, on better technologies of oil exploration, on finding alternative sources of energy, and on energy-saving measures. Compared with the difficulties of global oil markets, the situation with phosphate rock mining seems to be even more alarming, for the following reasons:

- There is no substitute for phosphate rock among all natural minerals.
- There are no alternatives to phosphate fertilizers in agriculture.
- It is not possible to reduce the P consumption either for people (1.0–1.5 g P per day) or for animals (5.0–50.0 g P per 100 kg weight per day).
- There are quite a few prospective reserves of phosphate rock currently under exploration.

Similarly to the case for oil, this forecast on peak phosphorus does not mean that by 2030 there will be no phosphate rock available on the Earth. It simply means that by that time the exploration and mining of remaining reserves will be much more costly and relevant global prices will grow enormously.

The publication of the new evaluation of URR as 12,000 Mt (von Horn and Sartorius 2009) instead of 8,000 Mt, as suggested earlier, provides only a small remedy because phosphate production does not seem to react on the warnings and continues to intensify.

In 2009, about 140 Mt phosphate rock was mined. With the strong growth of agrofuels, the amount of phosphate rock that needs to be mined will be forced to rise to about 171 Mt per annum in 2030, which means that in the time period until 2030 another 3,400 Mt phosphate rock would have to be mined. This amount is about half the amount of the reserves of the mines that are currently being operated.

How has the global community responded to the "Peak Phosphorus" warning? Very mildly, if at all. Comparing with the situation with oil, one can see substantial differences:

- In contrast to oil, it was confirmed that phosphorus has no alternatives.
- The global production of phosphates was not reduced.
- Ironically, strong measures on the promotion of biofuels drives higher phosphate mining rates.
- Most of the commercial P fertilizers remain water soluble.
- Sustainable agriculture and natural farming do not find adequate support.

Fortunately, in contrast to oil, phosphorus can be recycled. Constructive responses to the phosphorus peak include recreating a cycle of nutrients, for example, on a local basis, returning animal (including human) manure to cultivated soil as Asian people have done in the past. But the most promising is the scope of technologies enabling P recycling from sewage sludge, starting from the old (but unsafe) routine of sludge disposal on the fields and extending to the recently successful efforts of the complete recycling of environmentally friendly P fertilizers from sewage water streams, sludge, and incineration ash.

If this is not achieved by the middle of this century, phosphorus may demonstrate its alter ego.

—*Lucifer*

References

Deffeyes K (2005) Beyond oil—the view from Hubbert's peak. Hill and Wang, NY, USA
Dery P, Anderson B (2007) Peak phosphorus. Energy bulletin. Publication of the Post Carbon Institute, 13 Aug 2007
Khebab (2006) World production forecast. The oil drum. http://www.theoildrum.com/story/2006/11/13/225447/79. Accessed 14 Dec 2011
von Horn J, Sartorius C (2009) Impact of supply and demand on the price development of phosphate (fertilizer). In: Ashley K, Mavinic D (eds) International conference on nutrient recovery from wastewater streams. Fraunhofer Institute Systems and Innovation Research, Karlsruhe

Chapter 10
Phosphate Recycling or Welcome from Lucifer?

One of the most prominent thinkers of the twentieth century said: "Life can multiply until all the phosphorus has gone and then there is an inexorable halt which nothing can prevent." (Asimov 1974). The currently functioning phosphate supply chain, starting from its initial point (rock mining) and ending at the demand points (human and animal food), has many dissipation spots where masses of phosphate leave the supply chain and are scattered in the environment as pollutants.

Some dissipation spots are these:

- Phosphate mining sites
- Fertilizer production sites and transportation routes
- Land fertilization
- Livestock and food production
- Food processing
- Food consumption by humans and animals.

This chain is illustrated schematically in Fig. 4.1.

One of the few ways to decrease permanent loss of available phosphorus is a recovery of different forms of phosphate at these dissipation spots. Although in many references this activity is called "phosphorus recycling," it would be more correct to call it "phosphate recycling," or P recycling.

Let us analyze technical means that are available for P recycling from the following waste streams:

- Direct discharge from the phosphate mines
- Waste from fertilizer production plants
- Municipal sewage discharge
- Waste from agricultural and related activities: runoff from fields, wastes from animal/poultry farms and slaughterhouses (processing of cadavers, bones, etc.);

Such waste sources as phosphate mines, fertilizer production plants, slaughterhouses, and municipal sewage systems are called "point sources" because their waste streams are, or can be, concentrated and combined for the purpose of

M. Butusov and A. Jernelöv, *Phosphorus: An Element that could have been called Lucifer*, 89
SpringerBriefs in Environmental Science 9, DOI 10.1007/978-1-4614-6803-5_10,
© Mikhail Butusov and Arne Jernelöv, 2013

P-recycling. The runoff from fields, from habitats without functional sewage systems, and from many farms are called "diffuse (or "non-point") "sources."

In general it is much harder to handle the waste streams from diffuse sources than from the point ones. What can be done to minimize these losses and recycle phosphates from these waste streams?

The EU guidelines recommend the following universal approach with respect to the waste treatment generated at any of these locations:

1. Waste avoidance. If not possible, then:
2. Material utilization, meaning recycling and using the main valuable substance, phosphate. If not possible, then:
3. Thermal utilization, meaning combustion of the waste organic components and utilization of the generated thermal energy. If not possible, then:
4. Disposal in landfills.

10.1   Point Sources

10.1.1   Waste Streams of Phosphate Mines and Processing Plants

Starting from the P-rock mining sites and fertilizer plants, one should consider that the rock phosphate, as the main raw material for their operations, contains not more than ~30 % of the useful substance P_2O_5. If the producer intends first to convert the P rock into phosphoric acid, the rest of the mined mass is a waste. Therefore, the first option, that is, "waste avoidance," is impossible. Phosphate mines and fertilizer plants try to follow the second and fourth routes, "material utilization" or "landfilling." The degrees to which extent these routes are followed depend upon a concrete producer, whereas the range of practical solutions is not very wide—"from nothing to something."

The Environmental Protection Agency (USA) initiated a comprehensive research in Florida focused on three possibilities of potential use of the main waste from phosphoric acid production, phosphogypsum (PG): material for road construction, remediation of landfills, and agricultural uses. The findings were as follows:

- Application of PG as an additive for the construction of local roads was initially justified because PG is a cement-like material that strengthens over time. Indeed, several test roads were built and observed. It was found that, after several years, they were in better shape than when first built. The EPA, however, did not permit this application of PG except for temporary local service roads.
- The second suggestion was to scatter PG over existing waste landfills for remediation purposes. Laboratory studies indicated that PG contains such nutrients as calcium and sulfur that may enhance anaerobic bacterial activities and therefore increase the rate of waste decomposition. However, practical implementation of this idea would create huge dust clouds all over Florida, given the increasing

activity of tornados there. Several counties of Florida rejected the use of PG on their landfills;

- Finally, use of PG in agriculture was justified because this material contains large amounts of sulfur and can be brought into soils that are deficient in sulfur. The reason is that PG is the least expensive sulfur carrier in the world. Adding sulfur to lands used for grazing cattle should increase the protein content in grass and make it more digestible for two-stomach animals, thus achieving a 20 % weight gain on the improved diet. Rothamsted Research, the world's oldest agricultural research center, was assigned to determine what happens to various components of PG, including radionuclides, heavy metals, and nutrients. This research is in the midway stage and so far no conclusions can be drawn.

Recently some studies indicated that PG can be a potential resource of rare metals. The Russian company Phosagro (Phosagro 2012) together with the Belgian company Prayon announced their plan to launch, in 2012, a demonstration plant for processing of 25,000 t PG per year.

10.1.2 Waste from Slaughterhouses

Slaughterhouses are a part of the phosphate-bearing cycle attached to the "FOOD" areas on Fig. 4.1. The main P-containing product from the slaughterhouses is a "raw bone meal," a mixture of crushed and coarsely ground bones that can be used as an organic fertilizer for plants and was also formerly used as an animal feedstock. It contains about 22 % phosphoric acid and 4 % nitrogen. As a slow-release fertilizer, bone meal could be a good source of phosphorus. Probably the first deliberate use of phosphorus fertilizer began in Biblical times. People noticed that the best grapes grew in old battlefields. The most productive vineyards were planted over the unmarked graves of the fallen and fertilized by their bones. Bones and bone meal continued to be used as a phosphorus fertilizer by farmers through the ages and are still used today by organic farmers.

As a rule, bones are first chemically treated and then boiled, or treated with high-pressure steam to remove fats and gelatin. After this the residue is ground to the so-called bone meal that can be used as a fertilizer. The fineness of this product has a strong influence on the rapidity of its decay in the soil. Being more spongy and soft, it yields its phosphoric acid to the soil in a much shorter time than the hard "raw bone" produced from the untreated bones. Steaming also reduces the nitrogen content to about 1 % whereas the proportion of phosphoric acid is raised to 27 % or 28 %. Additionally, bone meal may also reduce the bioavailability of lead in soils contaminated with lead.

The main problem with the application of all types of bone meal is that, as of the 1990s, bone meal was identified as a trigger for bovine spongiform encephalopathy (BSE, or "mad cow disease") among livestock. It was believed that the bone meal was produced in the 1970s from the corpses of sheep or cows contaminated with

BSE. Since then, many certification bodies in Canada and Europe first prohibited using the bone meal for feeding the cattle. Then, this concern also spread among farmers who were using the bone meal as a fertilizer. As a result, the demand for this type of fertilizer declined.

The only way to destroy the organic contaminants in bone meal is its combustion. Bone meal contains enough organic substances to burn down, in presence of free oxygen, and create a pure ash containing up to 40 % P_2O_5. Bone ash is a white, powdery material primarily composed of calcium phosphate. It can be used as a fertilizer, but because its chemical composition (tricalcium phosphate) is more stable than that of organic bone meal, such application can be recommended only in special cases. In the meantime, bone ash found solid applications as an additive for making baking powder, polishing compounds, ceramics (such as a high-quality bone china), and pigments.

Therefore, with respect to bone meal, one can say that the combination of the second and third recommended EU options, "thermal and material utilization," was well developed.

10.1.3   Municipal Sewage Discharge

Municipal wastewater treatment is an essential public service indispensable for sustainable city management. Every city resident discharges daily to the municipal sewage systems between 0.3 and 0.7 m^3 dirty water per day. Although this water carries rather small amounts of contaminants (both organic and inorganic), the typical urban discharge contains 5–15 times more suspended solids and has up to 30 times higher "biochemical oxygen demand" (BOD) than is allowed for water streams to be returned to nature. BOD is a measure of the amount of oxygen dissolved in the water discharge that supports the microbial metabolism of organic compounds existing in native water reservoirs. This dissolved oxygen, after entering the environment, activates a complete oxidation of the organic compounds through generations of microbial growth, their death and decay, and thus suppresses the indigenous aquatic flora and fauna.

There are many methods and processes to treat wastewater. The most common approach uses three stages.

- Primary treatment (screening and clarification) to remove coarse solids: primary treatment involves screening, grinding, and sedimentation/clarification, to remove the floating and settleable solids found in raw wastewater. When raw wastewater enters the treatment plant, it is typically coarse screened to remove large objects, ground to reduce the size of the remaining solids, and then flows to primary sedimentation tanks. The sedimentation tanks provide sufficient capacity to establish quiescence in the wastewater, allowing solids with a higher specific gravity than water to settle and those with a lower specific gravity to float. Well-designed and well-operated primary treatment should remove 50–70 % of the suspended solids and 25–40 % of the BOD. Free oil, grease, and other floating substances are removed by skimmers from the surface of the primary sedimentation tanks. Typical retention time in the primary sedimentation tanks is 1.5 to 2.5 h. Chemical flocculants/polymers are frequently added to the primary sedimentation tanks to increase solids removal. Solids removed during primary treatment are dewatered and disposed of as part of the sludge treatment.

- Secondary treatment necessary to further reduce organic pollutants is based on aerobic digestion. Effluent from primary treatment is treated in large reactors or basins. In these reactors, an aerobic bacterial culture (the activated sludge) is maintained, suspended in the liquid contents. The bacteria metabolize the organic carbon in the wastewater, producing carbon dioxide, nitrogen compounds, and a biological sludge. Hydraulic retention time in the secondary reactors ranges from 6 to 8 h. Secondary treatment typically removes 70–85 % of the BOD entering with the primary effluent. Treated effluent from the aeration basins flows to secondary clarification. A portion of the sludge from the clarifier is recycled to the aeration basins/reactors and the rest is withdrawn. The waste sludge is dewatered and disposed of by various methods. This type of secondary treatment is now the most common, although it is not the most energy efficient. Recently, however, use of aerated lagoons became unacceptable because they required a too large land area.
- Tertiary treatment (also known as "advanced wastewater treatment") is becoming more common as the European discharge permits require removal of specific contaminants not normally extracted during conventional secondary treatment, first of all, the dissolved nutrients (nitrogen and phosphorus) that, after having been released to nature, encourage eutrophication (see Chap. 7).

There is enormous variability from plant to plant in wastewater flow rates (up to 1,000 Ml (million liters)/day), concentration of contaminants (depending upon the state of water supply pipes, collectors, etc.), type of processes used, the discharge regulations the effluent must meet, disinfection methods used, and, finally, the seasonal peaks, such as melting snow flows, that the plant must treat. In general, the more thorough is the waste treatment process, the richer in P is the resulting sludge.

A properly functioning wastewater treatment plant (WWTP) generates about 30–40 kg dry sludge per city resident per year. European plants generate more than 9 Mt dry sludge annually: the volume has a permanent tendency to increase so long as urbanization continues. This sludge contains about 1 M tons of phosphorus (as P_2O_5)—an amount sufficient for the fertilization of hundreds of thousands of hectares of agricultural land. The lack of standardization and country-specific regulatory requirements (even within the EU) makes it impractical to establish an optimal wastewater treatment and phosphate-recycling concept in terms of the system configuration and performance. It is clear, however, that the waste generated in such amounts should be one of the prime targets wherein economically and environmentally viable P-recycling technologies can help.

10.1.4   Sludge Management Options

WWTP sludge is a material consisting of three components:

• The nontoxic organic component, which represents about 55–60 % of the dry sludge (DS) weight
• The nontoxic mineral component, including phosphorus, which is almost 40 %
• The toxic organic and mineral component, such as pathogens, organic toxicants, and heavy metals (HM): the smallest but significant part deserving special attention

Main sludge management routes, currently in use or at the final development phase, are shown on Table 10.1 in consideration of their costs and of how efficiently they use material (organic and mineral) components and reduce the influence of toxicants (European Community 2002).

This table allows drawing several conclusions:

1. Although municipal sewage systems worldwide pollute the environment with smaller phosphorus-containing flows than runoff from fields and farms and even from fertilizer production and transportation operations, there are many possible options of how to reduce sludge pollution or even to recycle energy and part of P from these flows.
2. The scope of these options covers three of four EU-recommended treatments: landfilling, thermal utilization, and material utilization. The first treatment, waste avoidance, is not possible. Moreover, as the amount of WWTPs and quality of water purification processes grow, the amount of generated sludge also increases.
3. In several EU countries sludge is still routed from the sewage systems of some coastal towns directly to the sea despite the fact that this contradicts EU regulations.
4. There is no evident favorite among the nine most prospective or currently used sludge maintenance options. All options have their advantages and shortcomings. The choice of each municipality depends upon many factors: economic, political, and environmental.

Accordingly to capacities and national regulations/preferences, different countries develop different management preferences. Data at the brink of the twenty-first century show that in Austria more than 30 % of sludge was incinerated, 12 % was spread on the fields, and the rest (about 58 %) was landfilled (Juniper 2000). In Sweden and in Greece incineration was totally absent, while Sweden landfilled and field-spread sludge in equal shares and Greece landfilled 95 % of sludge. Ireland and UK indicated that "other" maintenance methods were also used (55 % and 30 %, respectively), which presumably was routing sludge streams to the sea or other aquatic sinks.

As to future trends, they do not look very clear. In 2020, Austria still plans to increase the proportion of incinerated sludge from the current 40 % to 85 %, Hungary, from 5% to 30 %, Slovakia, from 5% to 40 %, and Slovenia, from 25 % to 70 %; the Netherlands and Belgium will reach almost 100 %. On the other hand,

Table 10.1 Main options for processing and reuse of sewage sludge (costs of processes: average for ten European countries)

	Option	Advantages	Shortcomings	Criterion 1	Criterion 2	Criterion 3	Cost, €/t dry sludge (DS)
1	Reuse in agriculture	Fertilizer substitution Rather stable market	Legal barriers Concerns of users and retailers Needs drying and compacting	No	Yes	No	150–400
2	Composting	Simple Low capital cost	Requires large land Slow process Produces humus Marketing problems	Partly	Yes	No	250–600
3	Stabilization	Low cost Disinfects before use Can recycle cement kiln dust	Needs stable agro-consumer Needs cement producer nearby	No	No	HMs remain in product	200–400
4	Anaerobic digestion	Produces energy Produces compost Well proven	Energy conversion inefficient Humus marketing	Yes	Yes	No	300–800
5	Landfilling	Simple Low cost Can be secure route	Worst case exit Methane emissions Danger to nature Gas collection hard	No	No	No	200–600
6	Co-combustion	Additional fuel for cement kilns	Only small volumes can be used High costs Eco-impact	Yes	No	No	

(continued)

Table 10.1 (continued)

	Option	Advantages	Shortcomings	Criterion 1	Criterion 2	Criterion 3	Cost, ε/t dry sludge (DS)
7	Incineration	Well proven	Capital costs high	Yes	No	HMs remain in ash	450–800
		Secure end disposal	Eco-control strong				
		Generates energy	Ash, toxic residue				
			Public resistance				
8	Gasification/pyrolysis	Produces syn-gas	Capital costs high	Yes	No	HMs remain in residues	450–600
		Generates energy	Not proven enough				
		Low energy needed	Complicated				
		Lower emissions	Tar to be treated				
9	Incineration+	Generates energy	Capital costs high	Yes	Yes	Yes	600–1,000
	ASH DEC technology	Produces fertilizer	Not proven enough				
		Small residue amount	Complicated				

Criterion 1, utilization of organic part
Criterion 2, utilization of phosphorus
Criterion 3, removal of toxicants

France, UK, Sweden, the Czech Republic, and Ireland will remain the champions in the "agricultural outlet" with 60 % or more of sludge to be used on the fields.

Some countries are well informed about the risks of sludge land use but do not have adequate finance to install modern sludge treatment technologies. They still route a large part of the sludge to landfills: Greece, 55 %, Romania, 30 %, and Malta, 90 %. In this situation EU cannot formulate a common policy with regard to sludge management

Returning to Table 10.1, one should note that most of the parameters are qualitative and cannot be used for simple rating. For example, estimated costs of sludge incineration are more than €450/t DS, a figure that should discourage anyone interested in the implementation of this option. But this figure does not take into consideration that during incineration of 1 t sludge more than 2 MWh energy can be generated, and the remaining ash (in the amount of 450 kg of the initial sludge DS) contains P_2O_5 equivalent to 400 kg superphosphate (which costs €100–200/t). If a potential of sludge both as a fuel and resource of P fertilizer is included in cost evaluations, incineration of sludge could look more advantageous. ASH DEC technology used this possibility and allowed producing, from the sludge incineration ash, a slow-release P-based fertilizer, Phoskraft™, for agriculture and forestry (BAFU 2009).

In general, competition in future will develop between two options: thermal treatment and agricultural use. Promoters of thermal treatment emphasize the strengths of their choice:

- Possibility to generate "green energy"
- Reduction of transport costs
- Continuity of incineration treatment as opposed to the "seasonal" peaks inevitable in land use
- Better chances to recycle phosphorus from the incineration ash than from the sludge

At the same time, thermal treatment still has serious disadvantages:

- Costs of installations
- Emission control
- Disposal of ash or other residues (so long as it is not used for P recycling)
- Public perception

Proponents of land use operate with the following arguments in their favor:

- Political support: some politicians and scientists appeal that the land use is indeed a direct implementation of the second recommended EU priority ("material utilization")
- Adopting the routine "low-cost" attitudes in many countries
- Relative operational easiness
- No requirements for further investments
- Possibility of partial phosphorus release as a soil nutrient
- Possibility of using a part of the organic sludge content for soil remediation

The weaknesses of this approach include these:

* Presence of xenobiota (bacterial and other contaminants)
* Variable demand, that is, the necessity to provide a large storage space for sludge between agricultural seasons
* Large distances between production (wastewater plants) and application (fields) sites and complicated logistics thereof
* Growing resistance from farmers and food chain operators toward increasing soil pollution by inorganic (heavy metals) and organic contaminants. Indeed, for proper fertilization one needs to apply about 100 kg P_2O_5 (about 500 kg superphosphate) per hectare. There is up to 9 % of P_2O_5 in the best dry sludge, which means that 1 ha needs about 1,000 kg sludge. This amount contains (for Moscow sludge) 35 g Pb, 4.4 g Cd, 160 g Cu, 800 g Zn, and 420 g Cr. The presence of heavy metals in the same amount of the average Italian sludge is Pb, 221 g, As, 6.5 g, Hg, 0.98 g, Cd, 3.9 g, Ni, 76.2 g, Cr, 76.3 g, Zn, 1,228 g, and Cu, 367 g. For German sludge average data are 60 g Pb, 1.5 g Cd, 280 g Cu, 840 g Zn, and 53 g Cr. These annual sludge inputs for some metals exceed the upper limits of soil contamination. For example, in Sweden the upper limits of heavy metals in the year 2000 were established as such: 25 g Pb, 0.75 g Cd, 300 g Cu, 600 g Zn, and 40 g Cr. So, if the average Italian or German sludge is spread on the Swedish fields, this way of treatment automatically is illegal.

Nevertheless, the "pragmatic" supporters of the traditional sludge application routes in EU maintain that some findings indicate that much smaller concentrations of heavy metals were found in certain fields that have been consistently fertilized by sludge. They disregard that the chemical compounds of some metals tend to accumulate in soils, whereas others have increased solubility and mobility and therefore they are more easily taken up by the crops or migrate to the lower soil layers and to the surrounding environment.

What is the reaction of consumers to these discussions? In general, food consumers and retailers as well as farmers have become increasingly aware of the damage of inorganic contaminants and pathogens. When future quality standards for sludge and the receiving environment become more and more stringent, the "agricultural outlet," most popular during the past decades, may become invalid. Thus, the water utilities responsible for sludge routing need to urgently look for solutions that offer them greater operational and financial security.

10.1.5   P Recycling from Sewage Water Streams

There is a possibility of P recycling at the WWTP site before creation of sludge, that is, directly from the sewage water where phosphorus can be found in particulate, colloidal, and dissolved forms.

The dissolved fraction consists of the following:

1. Organically bound phosphates, including traces of P from detergents
2. Polyphosphates, also stemming from washing compounds, such as those used in dishwashers
3. Orthophosphates stemming from urine, which represent the major part of the dissolved fraction

P recycling directly from the wastewater before or even instead of its conventional treatment can be an attractive alternative. Two most interesting technologies were developed for nonthermal P recycling from the sewage stream.

10.1.5.1 Technology Developed by Ostara, Inc. (Vancouver, Canada) and Unitika Ltd. (Osaka, Japan)

Technology developed by Ostara, Inc. (Vancouver, Canada) and by Unitika Ltd. (Osaka, Japan) allows separating, from the stream processed in the water treatment plant, about 30–50 % of contained phosphorus by means of precipitation. Precipitation results in fine crystals of struvite, a water-soluble magnesium–ammonium phosphate having a typical concentration of P_2O_5 12.5 %, Mg 9.5 %, and N 5.3 %.

The advantage of the struvite production process is that the P-containing product shows only traces of heavy metals and can be used as a pure phosphate for the food and cosmetics industries as well as for manufacturing high-quality fertilizers. However, so far the achieved P-recovery rate is not above 50 %, and therefore the downstream flow from the sewage needs to be treated as usual.

10.1.5.2 Technology of DHV Water BV

Technology developed by DHV Water BV (Amersfoort, The Netherlands) and named Crystalactor offers the use of crystallization instead of precipitation to receive dewatered P-containing material. By dosing calcium or magnesium salts (e.g., lime, calcium chloride, magnesium hydroxide, or magnesium chloride) to the water, the solubility parameters typical for, respectively, calcium phosphate (CP), magnesium phosphate (MP), or other salts of phosphoric acid are exceeded and subsequently phosphate is transformed from the aqueous solution into a solid crystal material in the form of pellets. The primary difference from conventional precipitation is that in the Crystalactor the transformation process is well controlled and that pellets with a typical size of about 1 mm with water content less than 5 % are produced, whereas in the dewatered sludge phosphates are finely dispersed, microscopic sludge particles with a water content greater than 30 %.

Pellets can be used for several applications:

- Raw material for the production of phosphoric acid
- Intermediate product for conventional fertilizers
- Direct use as a slow-release fertilizer

An advantage of both technologies is that if the conventional dewatering process is installed downstream, the received organic residue can be used for combustion, thus allowing generation of both P-containing mineral fraction and organic residue for thermal generation. A common positive feature of both nonthermal technologies is also that they can be built in the operational wastewater treatment plants (WWTP) as independent modules and, resulting in additional commercial products, can improve the profits of these plants. However, in both cases either the mineral or the organic fraction accumulates most of the heavy metals from the wastewater, and special studies are needed before recommending these methods for processing the sewage water of most industrial cities.

10.2   Recycling from Agricultural Activities

Water pollution caused by poultry and animal farms was discussed in Chap. 5. Despite multiple studies in this area, especially in Baltic countries, on how to economically convert the manure stored at and spilled from these farms, practical results are still to be expected.

As to P recycling from field runoff, because of overuse of water-soluble fertilizers, especially in the poor soils of northern countries, these fields together with farms remain among the badly controlled "non-point" sources of pollution, a major environmental problem not only for the Baltic Sea (Helcom 2011).

Apart from technical complexities in allocating the runoff paths, there are purely methodological problems:

- Phosphorus substances leaching from the fields are hard to identify (colloidal, dissolved, associated with other chemical forms) and to assign with an exact toxicity grade
- So far, only some rather complicated recommendations were made on sustainable soil management that may help reduce runoff and catch more nutrients in the productive layer
- In general, so long as water-soluble fertilizers are commonly used, the runoff cannot be controlled and measures recommended by EU for waste management practices cannot be imposed.

The only substantial improvement in P recycling from multiple waste streams was achieved only with respect to discharges from municipal sewage systems.

References

Asimov I (1974) Asimov on chemistry. Doubleday Publishing, NY, USA, LCCN 73–15322. ISBN ISBN 0-385-04100-4

BAFU (2009) Ruekgewinnung von Phosphor aus der Abwassereinigung. BAFU, Switzerland

European Community (2002) EC report. "Disposal and recycling routs for sewage sludge" part 4 "Economic report" 2002 ISBN 92-894-1801-X. http://ec.europa.eu/environment/waste/sludge/pdf/sludge_disposal4.pdf. Accessed 18 Dec 2011

Helcom (2011) Fifth Baltic Sea pollution load compilation by Helsinki commission. In: Baltic Sea environment proceedings no. 128. http://www.helcom.fi/stc/files/Publications/Proceedings/BSEP128.pdf. Accessed 18 June 2012

Juniper (2000) Pyrolysis and gasification of waste. Juniper Consultancy Services

Phosagro (2012) Phosagro will extract rare metals (in Russian). RBC Daily 29 Oct 2012. http://www.rbcdaily.ru/2012/10/29/industry/562949985020675. Accessed 22 Oct 2012